On Superconductivity and Superfluidity

Vitaly L. Ginzburg

On Superconductivity and Superfluidity

A Scientific Autobiography

With 25 Figures

 Springer

Professor Vitaly L. Ginzburg
Russian Academy of Sciences
P.N. Lebedev Physical Institute
Leninsky Prospect 53, 119991 Moscow, Russia

Translation from the Russian original version
O sverkhprovodimosti i o sverkhtekuchesti. Avtobiografia, by V.L. Ginzburg
(Moskva: Izdatel'stvo Fiziko-matematicheskoi literatury, 2006)

ISBN 978-3-642-08761-5   e-ISBN 978-3-540-68008-6

Cover design: WMX Design GmbH, Heidelberg

Printed on acid-free paper

9 8 7 6 5 4 3 2 1

springer.com

Preface to the English Edition

The English version of the book does not differ essentially from the Russian version[1]. Along with a few notes and new references I included Part II to Article 3 and added some new materials to the 'Nobel' autobiography. Furthermore, Article 7 (M. Cardona and W. Marx "Vitaly L. Ginzburg – a bibliometric study"), which was published in *Journal of Superconductivity and Novel Magnetism*, v. 19, No. 3–5, July 2006 is included as an appendix.

My special thanks are due to Prof. Manuel Cardona and Prof. Werner Marx who kindly allowed publishing their paper as an appendix to this book (with some new minor author's amendments).

Also, I am grateful to M.S. Aksent'eva, E.A. Frimer, G.M. Krasnikova and S.G. Rudnev for their assistance in the preparation of the English manuscript.

Moscow, September 2008 *V.L. Ginzburg*

---

[1] V.L. Ginzburg, *O sverkhprovodimosti i o sverkhtekuchesti. Avtobiografia* (Moskva: Izdatel'styvo Fiziko-matematicheskoi literatury, 2006)

Preface to the Russian Edition

The Nobel Prize in Physics, 2003 was awarded to A.A. Abrikosov, A.J. Leggett and myself 'for pioneering contribution to the theory of superconductors and superfluids'. It does not mean that the contribution was made in joint works with these authors. Specifically, I do not have any joint publications with A.A. Abrikosov and A.J. Leggett.

In conformity with the established rules, I prepared the Nobel Lecture and delivered it in Stockholm, on December 8, 2003, and have also written the autobiography. The copyright for both these texts belongs to the Nobel Foundation, and they already appeared in English. As in the previous years, all 2003 Nobel Lectures with autobiographies were published in the book *Les Prix Nobel/The Nobel Prizes 2003* (Nobel Prizes, Presentations, Biographies, and Lectures). Moreover, lectures, usually without autobiographies, are published in different journals. All Nobel Lectures in Physics 2003 were published in Russian in the journal *Uspekhi Fizicheskikh Nauk* v. 174, No. 11 (2004) [English edition *Physics-Uspekhi* v. 47, No. 11 (2004)]. By the way, this journal always publishes the Nobel Lectures in Physics, and sometimes in other sciences as well. Even before the Nobel Foundation published the lectures, they kindly granted us the permission for placing my Lecture and autobiography in Russian on the site of *Physics-Uspekhi* www.ufn.ru.

It is the publication of these materials that is the main purpose of publishing this book. At the same time, it also contains some other related or auxiliary materials. For example, the article "Superconductivity and Superfluidity (What was Done and What was Not Done)" is a rather extensive survey of everything which I did in the field of superconductivity and superfluidity. This survey was written in 1997 and published in *Physics-Uspekhi* and later in my book "About Science, Myself and Others" (Moscow: Fizmatlit, 2003) [The English edition was published by the Institute of Physics Publishing in 2005). When it was prepared for publication in this book, some minor changes were brought into it, in comparison with the Russian publication of 2003. This survey is added by a small article "Several Notes on Studies of Superconductivity" (*Uspekhi Fizicheskikh Nauk* v. 175, p. 187, 2005) [*Physics-Uspekhi*

v. 48, p. 173 (2005)]. It is followed by my joint work with L.D. Landau, which was published in 1950 and occupies a special place in the studies in the field of the theory of superconductivity. The book is concluded by the autobiography written at the request of the Nobel Foundation ('the Nobel' autobiography) and the article 'Scientific Autobiography – an Attempt' published earlier in my book "On Physics and Astrophysics" (1992, 1995) [The English edition under the title "The Physics of a Lifetime" was published by Springer in 2001]. Publishing this article here seems appropriate to me, since in the 'Nobel' autobiography I do not at all touch on the contents of my scientific works. At the end of the articles there are comments prepared for the present edition (the references to them are marked in the text by Arabic figures with an asterisk above the line).

Taking advantage of this opportunity, I thank M.S. Aksent'eva, T.S. Vaisberg, S.V. Volkova and L.A. Paniushkina for their assistance in the preparation of this manuscript.

Moscow, September, 2005   *V.L. Ginzburg*

Contents

1

On Superconductivity and Superfluidity (What I Have and Have Not Managed to Do), As Well As on the 'Physical Minimum' at the Beginning of the 21st Century[1*]

(Nobel Lecture in Physics. December 8, 2003)

1.1 Introduction

First of all, I would like to express my heartfelt gratitude to the Royal Swedish Academy of Sciences and its Nobel Committee for physics for awarding me the 2003 Nobel Prize in Physics. I am well aware of how difficult it is to select no more than three laureates out of the far greater number of nominees. So, this award is the more valuable.

Personally, I have two additional motives for appreciating the award of the Prize. Firstly, I am already 87, the Nobel Prize is not awarded posthumously, and posthumous recognition is not all that so significant to me since I am an atheist. Secondly, the 1958 and 1962 Nobel Prizes in Physics were awarded, respectively, to Igor' Evgen'evich Tamm and Lev Davidovich Landau. Outside of high school, the notion of a teacher is very relative and is quite often applied by formal criteria: for instance, it is applied to the supervisor in the preparation of a thesis. But I believe that the *real* title of teacher can be given only appropriately to those who have made the greatest impact on ones work and whose example one has followed. Tamm and Landau were precisely these kind of people for me. I feel particularly pleased, because in a sense I have justified their good attitude toward me. Of course, the reason lies not with the Prize itself, but with the fact that my receiving the award after them signifies following their path.

Now about the Nobel Lecture. It is the custom – I do not know whether by rule or natural tradition – that the Nobel Lecture is concerned with the work for which the Prize was awarded. However, I am aware of at least one exception. P.L. Kapitza was awarded the 1978 Prize for "his basic inventions and discoveries in the area of low-temperature physics." However, Kapitza's Lecture was entitled "Plasma and the Controlled Thermonuclear Reactions." He justified his choice of the topic as follows: he had worked in the field of low-temperature physics many years before he had been awarded the Prize and he believed it would be more interesting to speak of what he was currently engaged in. That is why P.L. Kapitza spoke of his efforts to develop a fusion reactor employing high-frequency electromagnetic fields. By the way, this path has not led to success, which is insignificant in the present context.

I have not forgotten my "pioneering contributions to the theory of super-conductors and superfluids" for which I have received the Prize, but I would rather not to dwell on them. The point is that in 1997 I decided to sum up my activities in the corresponding field, and I wrote a paper entitled "Supercon-ductivity and Superfluidity (What Was Done and What Was Not Done)" [1,2]. In particular, this article set out in detail the story of quasi-phenomenological superconductivity theory constructed jointly with Landau [3]. Under the cir-cumstances, it would be unnecessary, and above all, tedious, to repeat all that. Furthermore, the Ginzburg–Landau theory of superconductivity, which I call the Ψ-theory of superconductivity, is employed in the work of A.A. Abrikosov [4], and he will supposedly dwell on it in his Nobel Lecture. This is to say nothing of the fact that the Ψ-theory of superconductivity has been covered in many books (see, for instance, $[5,6]^{2^*}$). At the same time, there are several problems bearing on the field of superconductivity and superfluidity which I have taken up and which have not been adequately investigated. This is why I have decided to dwell on these two most important problems in my lecture.

The case in point is thermoelectric effects in the superconducting state and the Ψ-theory of superfluidity. However, before I turn to these issues, I will cover briefly the entire story of my activities in the field of superconductivity. At the end of the lecture I will allow myself to touch on some educational program for physicists (the issue of a 'physical minimum'), which has been of interest to me for more than thirty years.

1.2 A Brief Account of My Activity in the Field of Superconductivity, Prior to the Advent of High-Temperature Superconductors

L.D. Landau was in prison for exactly one year and was released on April 28, 1939, primarily due to the efforts of P.L. Kapitza, who became his 'personal guarantee'.[1] Landau resided in this state until his premature death in 1968. The Landau 'case' was officially discharged by virtue of "*corpus delicti*" ("ab-sence of a basis of a crime") only in 1990 (!). The imprisonment had a strong effect on Landau, but fortunately it did not bereave him of his outstanding capabilities as a physicist. That is why he 'justified the confidence', as they said at that time, of those who released him on bail instead of shooting him or leaving him to rot in jail (Landau personally told me that he had not been far from death) by constructing his superfluidity theory [7]. I was present at his report on this topic in 1940 or maybe in the beginning of 1941 (the paper was submitted for publication on May 15, 1941). Also considered at the end

---

[1] For more details, see for example Article 10 in [2].

Fig. 1.1. I.E. Tamm (*left*), L.D. Landau (*right*)

of this paper was superconductivity, which was treated as the superfluidity of electron liquid in metals.

That work impressed me, of course, but at that time I was enthusiastic about quite a different range of questions, namely, the theory of higher spin particles. That is why, I did not take up the low temperature subject right away. Shortly afterwards our lives radically changed when the war broke out (for the USSR it began on June 22, 1941). The Physical Institute of the USSR Academy of Sciences, where I was working and still work, was evacuated from Moscow to the town of Kazan, where many difficulties were encountered, which I describe in my autobiography. In any case, it was not until 1943 that I made an attempt, in the spirit of the Landau theory of superfluidity [7], to do something applied to superconductivity.[2] That work [9] is of no great value today, but I believe there were some interesting points in it, for Bardeen considered it at length in his famous review [10]. Even at that time I was aware the work was poor, and therefore, did not submit it to a journal in English, which we would normally have done at that time (the journal – *Journal of Physics USSR* – was terminated in 1947 during the 'cold war'). My next paper was concerned with thermoelectric effects in the superconducting state [11], and its destiny seems to be unusual and strange. The point is that 60 years have passed, but some predictions made in that work have never been verified and thermoelectric effects in the superconducting state have not been

---

[2] It is true that somewhat earlier I had considered the problem of light scattering in Helium II [8] on the basis of the Landau theory [7].

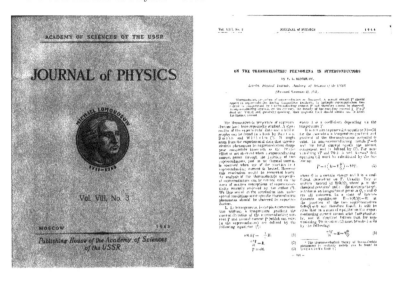

Fig. 1.2. *Journal of Physics*, published in English in the Soviet Union

adequately investigated. I myself returned to these problems more than once, but made no significant progress. Appeals addressed to other physicists have only a minor effect, for the problem is out of fashion. Here I would like to take advantage of my last opportunity to bring it to the attention of physicists; Sect. 1.4 is concerned with this question.

However, the question of thermoelectric effects in superconductors, while interesting, is still a particular problem, which evidently emerges only in the presence of a temperature gradient. Furthermore, at that time there existed no thorough theory of superconductivity, even under thermodynamic equilibrium. The fact is that the well-known London theory advanced in 1935 [12] (also discussed in Sect. 1.4 of this lecture) yielded much, and is widely employed under certain conditions even nowadays [5, 6, 13], but it is absolutely insufficient. The last-mentioned circumstance was largely elucidated in my next work, performed as far back as 1944 [14]. Specifically, the London theory is inapplicable in a strong magnetic field (in the theory of superconductivity, the field is termed strong when it is on the order of the critical magnetic field H_c; we are dealing with type-I superconductors). From the London theory it follows also that the surface energy σ_{ns} at the interface between the normal and superconducting phases is negative, and to attain the positiveness σ_{ns} one is forced to baselessly introduce some additional and, moreover, high surface energy of non-electromagnetic origin. Therefore, it became evident that the London theory had to be generalized. This problem was solved in 1950 in the

Ψ-theory of superconductivity [3].[3] This brings up the question, which has been repeatedly addressed to me: why did it take five years after the work in [14], in which the necessity of generalizing the London theory was recognized, to construct the Ψ-theory? Of course, I cannot answer this question as regards other physicists. As for myself, to some extent I was nearing my objective, as described in [1]. However, I believe the main reason for the slowness of this process lay with the fact that I did not focus my attention on the theory of superconductivity. Theoretical physicists have the good fortune to be able to work almost simultaneously in different directions and, in general, to move from one subject to another. Specifically, in the period from 1944 to 1950, apart from superconductivity and superfluidity, I was engaged in radio wave propagation in the ionosphere (plasma), radio-frequency solar radiation, light scattering in liquids, the theory of transition radiation (I.M. Frank and I jointly called attention to the existence of this effect), the relativistic theory of higher-spin particles (in part, jointly, with Tamm), undulator radiation, the theory of ferroelectrics, and other things. Of special note is the fact that ferroelectric effects (as applied primarily to $BaTiO_3$) were considered [15] on the basis of the Landau theory of phase transitions, and this direction subsequently progressed (see Article 5 in [2]).[4]

The Ψ-theory of superconductivity [3] is, if you like, an application of the Landau theory of phase transitions to superconductivity. In this case, some scalar complex Ψ function fulfills the role of the order parameter. By virtue of the foregoing I restrict myself to giving the equations employed for Ψ and the vector electromagnetic field potential $\boldsymbol{A}$ (as is generally known, rot $\boldsymbol{A} = \boldsymbol{H}$, where $\boldsymbol{H}$ is the magnetic field strength, which does not differ from the magnetic induction $\boldsymbol{B}$ in this case; furthermore, advantage is taken of the gauge div $\boldsymbol{A} = 0$):

$$\frac{1}{2m^*}\left(-i\hbar\nabla - \frac{e^*}{c}\boldsymbol{A}\right)^2\Psi + \alpha\Psi + \beta\left|\Psi\right|^2\Psi = 0, \tag{1.1}$$

$$\Delta\boldsymbol{A} = -\frac{4\pi}{c}\boldsymbol{j}_\mathrm{s}, \quad \boldsymbol{j}_\mathrm{s} = -\frac{ie^*\hbar}{2m^*}\left(\Psi^*\nabla\Psi - \Psi\nabla\Psi^*\right) - \frac{(e^*)^2}{mc}\left|\Psi\right|^2\boldsymbol{A}. \tag{1.2}$$

We consider an equilibrium or, in any case, a stationary state, and we assume that the normal current density in the superconductor is $\boldsymbol{j}_\mathrm{n} = 0$ (the

---

[3] As already mentioned, this theory is commonly referred to as the Ginzburg–Landau theory. However, I resort to the term Ψ-theory of superconductivity, because it seems to me that using one's own name rings, at least in Russian, somewhat pretentiously. Furthermore, a similar theory, as applied to superfluidity, was jointly elaborated in my work not with Landau, but with L.P. Pitaevskii and A.A. Sobyanin.

[4] For more details on the abovementioned work and other works of mine, see the article "A Scientific Autobiography – An Attempt" in [16] and Chap. 6 in this book.

total current density is $\boldsymbol{j} = \boldsymbol{j}_{\mathrm{s}} + \boldsymbol{j}_{\mathrm{n}}$, where $\boldsymbol{j}_{\mathrm{s}}$ is the superconducting current density). Furthermore, at the superconductor-vacuum interface we impose the boundary condition:

$$\boldsymbol{n}\left(-\mathrm{i}\hbar\nabla - \frac{e^*}{c}\,\boldsymbol{A}\right)\Psi = 0, \tag{1.3}$$

where $\boldsymbol{n}$ is the normal to the interface.

In the vicinity of the critical temperature T_{c}, at which there occurs the normal-to-superconducting phase transition in the equilibrium case, in the Ψ-theory it can (and even must) be assumed that:

$$\alpha = \alpha'_{\mathrm{c}}(T - T_{\mathrm{c}}), \quad \beta = \beta(T_{\mathrm{c}}) \equiv \beta_{\mathrm{c}} > 0, \quad \alpha'_{\mathrm{c}} > 0 \tag{1.4}$$

and the superconductor behavior is determined by the parameters:

$$\delta_0 = \sqrt{\frac{m^* c^2 \beta_{\mathrm{c}}}{4\pi (e^*)^2 |\alpha|}}, \quad \varkappa = \frac{m^* c}{e^* \hbar} \sqrt{\frac{\beta_{\mathrm{c}}}{2\pi}} = \frac{\sqrt{2}\,e^*}{\hbar c} H_{\mathrm{cm}} \delta_0^2. \tag{1.5}$$

Here, δ_0 is the depth of penetration of the weak magnetic field $H \ll H_{\mathrm{cm}}$ and H_{cm} is the critical magnetic field for massive samples (earlier, mention was made of the critical field H_{c}, which, say, for films is stronger than H_{cm}).

Since the Ψ-theory is phenomenological, the values of mass m^* and charge e^* are beforehand unknown. In this case, since Ψ is not an observable quantity (among the observable quantities are, in particular, the δ_0 and H_{cm} quantities), the mass can be arbitrarily selected: it is not among the measurable (observable) quantities. The question of choice of the e^* value is very interesting and intriguing. It seemed to me from the outset that e^* is some effective charge, which may be different from the electron charge or, as is said on occasion, the free-electron charge e. However, Landau did not see why e^* should be different from e, and in our paper [3] it is written as a compromise that "there are no grounds to believe that the charge e^* is different from the electron charge." I retained my opinion and saw that the way to solve this question was to compare the theory with the experiment. Specifically, the charge e^* enters in expression (1.5) for $\varkappa$, where δ_0 and H_{cm} are measured by the experiment; at the same time, $\varkappa$ enters into the expression for the surface energy σ_{ns}, for the depth of penetration in the strong field (the field $H \gtrsim H_{\mathrm{cm}}$), and for the limiting fields of the overcooling and overheating of superconducting samples. Following the path of comparing the theory with the experiment, I arrived at the conclusion [17] that $e^* = (2-3)e$. When I discussed this result with Landau, he raised an objection, which he had evidently been guided by before, though had not advanced it. Specifically, with the charge e^* assumed to be an effective quantity such as, say, the effective mass m_{eff} in the theory of metals and semiconductors, the effective charge may and, generally speaking, will depend on the coordinates, because the parameters that characterize the semiconductor are functions of the temperature, the pressure, and the composition, which in turn may depend on the coordinates $\boldsymbol{r}$. If $e^*(\boldsymbol{r})$, the gauge (gradient) invariance of (1.1), (1.2) of the Ψ-theory is lost. I did not find objections to this remark, and in article [17] outlined the situation (reporting Landau's opinion, naturally, with his permission). The solution, however, was quite simple. After the advent of the Bardeen–Cooper–Schrieffer (BCS) theory in 1957 [18], it became clear that in superconductors there occurs 'pairing' of electrons with opposite momenta and spins (I imply the simplest case). The resultant 'pairs', which are sometimes referred to as the Cooper pairs, possess zero spin and are Bose particles or, to be more precise, quasi-particles. The Bose–Einstein condensation of these pairs is responsible for the origin of superconductivity. By the way, as early as 1952 I noted [19] that the charged Bose gas would behave like a superconductor, but did not arrive at the idea of pairing. Interestingly, it had been advanced [20, 21] even before Cooper [22]. It is immediately apparent from the BCS theory that the role of charge e^* in the theory of superconductivity should supposedly be played by the pair charge, i.e., $2e$. This fact was proved by L.P. Gor'kov [23], who derived the Ψ-theory equations from the BCS theory. Therefore, Landau was right in the sense that the charge e^* should be universal, and I was right in that it is not equal to e. However, the seemingly simple idea that both requirements are compatible and $e^* = 2e$ occurred to none of us. After the event one may be ashamed of this blindness, but this is by no means a rare occasion in science, and it is not that I am ashamed of this blindness, but am rather disappointed that it did take place.

Many results were obtained in our work [3]. For small values of the parameter $\varkappa$ we calculated the surface energy σ_{ns} and pointed out that it lowers with increasing $\varkappa$ and vanishes when $\varkappa = \varkappa_{\mathrm{c}} = 1/\sqrt{2}$. Relying on the available experimental data we believed that for pure superconductors $\varkappa < \varkappa_{\mathrm{c}}$, and this is generally correct. In any case, we considered in detail only the superconductors with $\varkappa < \varkappa_{\mathrm{c}}$, which now are termed type-I superconductors. Subsequently, I would also restrict myself to the investigation of type-I superconductors (a certain exception is [24]). In 1950, as well as previously, the superconducting alloys were known to usually behave in a significantly different manner than pure superconductors. Particularly clear data concerning alloys were obtained by L.V. Shubnikov[5] and his collaborators in Kharkov in the mid-1930s (see the references and the results in [25]; this material was also touched upon in [26]. For more details see [27]). In [27], use is made of the term 'Shubnikov phase' for the alloys investigated by Shubnikov. However, an understanding of the situation was lacking, and Landau and I, like many others, believed that alloys are an 'unsavory business', and did not take an interest in them, restricting ourselves to the materials with $\varkappa < \varkappa_{\mathrm{c}}$ for which $\sigma_{\mathrm{ns}} > 0$, i.e., type-I superconductors. True, as noted in A. Abrikosov's paper [4] and in [5], Landau hypothesized that alloys are the ones where $\varkappa > 1/\sqrt{2}$, i.e., they are type-II superconductors, according to present-day concepts.

---

[5] In 1937, when Stalin's terror was in full swing, L.V. Shubnikov was arrested and shot.

The solution of different problems on the basis of Ψ-theory equations was our concern in the bulk of our work [3]. Apart from the abovementioned question of the energy σ_{ns}, we considered primarily the behavior of superconducting plates and films in the external magnetic field, and, in some cases, in the presence of current, and in doing this compared the theory with the experiment. Subsequently, Landau took no interest in such calculations and, in general, in the development of the Ψ-theory. My own effort made in this direction is described in [1]. Here, I restrict myself to the mention of a fairly evident yet important generalization of the Ψ-theory [3], in which superconductors were assumed to be isotropic, to the anisotropic case [28]. Furthermore, investigations were made of the overheating and overcooling of superconductors in the magnetic field [29] and of the quantization of magnetic flux in the case of a superconducting cylinder with an arbitrary wall thickness [30], and the Ψ-theory was compared with the experiment after the construction of the BCS theory [31]. Of special note is [32], which was developed in [33], had little bearing on the Ψ-theory, and applied to ferromagnetic superconductors. Such superconductors had not been discovered by that time, and [32] put forward the explanation for this fact related to the inclusion of magnetic energy. Subsequently (after the construction of the BCS theory), it became clear that the emergence of superconductivity in ferromagnetics is also hampered due to spin interaction. I was not engaged in that problem, but would like to call attention to the following. Certain considerations were given in [32], which allowed changing the role of the magnetic factor (the use of thin films and materials with a relatively strong coercive force). I do not think that anyone has given any attention to these possibilities, for old papers are seldom read. Of course, I do not feel sure that at the present stage one can find something of interest in [32, 33] – I would just like these papers to be looked at.

Long ago, in 1943, I engaged in the study of superconductivity because at that time this phenomenon appeared to be the most mysterious one in the physics of the condensed state. However, after the construction of the Ψ-theory, and especially the BCS theory, the picture generally became clear as regards the materials known at that time. That is why I lost particular interest in superconductivity, though I worked in this area episodically (see, for instance, [30, 34]). My interest was rekindled in 1964 in connection with the formulation of the problem of the feasibility of high-temperature superconductors (HTSCs). Mercury – the first superconductor discovered in 1911 – possesses $T_c = 4.15$ K, while the boiling temperature of ^{4}He at atmospheric pressure is $T_{b, ^4He} = 4.2$ K. By the way, from 1908 to 1923, for fifteen long years, liquid Helium could be obtained only in Leiden, and low-temperature physics research was pursued on a very small scale, judged by present-day standards. For example, it suffices to note that the bibliography given at the end of monograph [26] contains about 450 references to the papers on superconductivity (or, sometimes, related problems) over the period from 1911 to 1944; among them, only 35 references fall within the 1911–1925 period. Meanwhile, after 1986–1987, when high-temperature superconductivity was discovered, during the 10 subsequent years approximately 50,000 papers were published, i.e., about 15 papers per day (!).

There can be no doubt that immediately after the discovery, and first investigations of, superconductivity, the question arose as to why this phenomenon is observed only at low temperatures or, in other words, Helium temperatures. Naturally, there was no way to provide the answer until the nature of superconductivity was understood, i.e., until the construction of the BCS theory in 1957 [18]. The following expression was derived for the critical temperature in this theory:

$$T_c = \theta \exp\left(-\frac{1}{\lambda_{\text{eff}}}\right), \tag{1.6}$$

where $k_B\theta$ is the energy range near the Fermi energy $E_F = k_B\theta_F$, in which the conduction electrons (more precisely, the corresponding quasi-particles) are attracted together, which is responsible for pair production and the instability of the normal state; furthermore, in the simplest case, $\lambda_{\text{eff}} = \lambda = N(0)V$, where $N(0)$ is the electronic level density near the Fermi surface in the normal state and V is some average matrix element of electron interaction which corresponds to the attraction. In the BCS theory, in its initial form, the 'coupling constant' λ_{eff} and, specifically, λ is assumed to be small ('weak coupling'), i.e.,

$$\lambda \ll 1. \tag{1.7}$$

As regards the temperature θ, in the BCS theory it was assumed that

$$\theta \sim \theta_D, \tag{1.8}$$

where θ_D is the Debye temperature of the metal, for the interelectron attraction was thought to be due to the electron-phonon interaction (as is generally known, the highest phonon energy in a solid is of the order of $k_B\theta_D$). Typically, $\theta_D \lesssim 500$ K and $\lambda \lesssim 1/3$; whence it follows, according to (1.6), that $T_c \lesssim 500 \exp(-3) = 25$ K or more generally:

$$T_c \lesssim 30\text{--}40 \text{ K}. \tag{1.9}$$

Defining all this more precisely would be out of place here. However, it seems to me that the aforesaid will suffice to understand why the condition (1.9) is fulfilled for typical metals, and even safely fulfilled. In particular, prior to the discovery of high-temperature superconductivity in 1986–1987, all attempts to discover or produce a superconductor with the highest possible critical temperature had led in 1973 to the production of only the Nb$_3$Ge compound with $T_c = 23$–24 K. Of course, in what follows I do not endeavor to find the exact values of various parameters; they depend on the purity and processing of samples, etc.

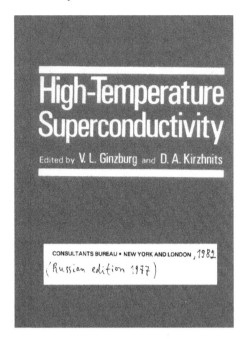

Fig. 1.3. Book *High-Temperature Superconductivity* (ed. by V.L. Ginzburg and D.A. Kirzhnits) published in Russian in 1977 and in English in 1982

1.3 On High-Temperature and Room-Temperature Superconductors (HTSCs and RTSCs)

The advent of the BCS theory made it possible to envisage the feasibility of a radical elevation of the critical temperature. It may be that I am not familiar with some facts, but to my knowledge this question was clearly and constructively posed for the first time by Little in 1964 [35]. Being forced to outline the following part of this section quite schematically, owing to lack of space, I can only mention that Little proposed considering the possibility of replacing the phonon mechanism of attraction between conduction electrons with the same attraction arising from the interaction with bound electrons present in the same system. I call this mechanism excitonic or electron-excitonic; to state it in plain terms, we are dealing with the replacement of phonons with excitons – excitations in the system of bound electrons. True, this term is not universally used in the literature. In his case, Little employed a quasi-one-dimensional model, in which some conducting 'spine' was surrounded by side 'polarizers', say, organic molecules. For electronic excitons or, in other words, for the excited states of bound electrons, the characteristic temperatures $\theta_{\mathrm{ex}} = E_{\mathrm{ex}}/k_{\mathrm{B}} \lesssim \theta_{\mathrm{F}} \sim 10^4$–$10^5$ K and, in any case, the values $\theta_{\mathrm{ex}} \sim 10^4$ K are quite realistic. It is, therefore, evident that replacing $\theta \sim \theta_{\mathrm{D}}$ in (1.6) with $\theta \sim \theta_{\mathrm{ex}}$ gives us the values $T_{\mathrm{c}} \lesssim 10^3$ K (when, say, $\lambda \sim 1/3$). Of course, these are no more than words, for it is still unclear how to realize the Little model, and this has never been accomplished. Furthermore, it became clear that the fluctuations in quasi-one-dimensional systems are so strong that the transition to the superconducting state is unlikely to occur. However, having familiarized myself with the paper [35], I put forward straight away [36] a quasi-two-dimensional model, wherein a plane conductor is in contact with a dielectric, say, a dielectric film. We termed the development of this version – the alternation of thin conducting layers with dielectric layers – a 'sandwich.' Going over from the quasi-one-dimensional model to the quasi-two-dimensional model was not accidental, for immediately before this work [36] D.A. Kirzhnits[6] and I had considered [37], not in connection with the high-temperature superconductivity problem, the problem of two-dimensional (surface) superconductivity. By the way, this problem is still of interest in itself, but I cannot enlarge on it for lack of space and will restrict myself to giving references [36–38].

Compared to quasi-one-dimensional systems, quasi-two-dimensional systems have the advantage that they exhibit significantly weaker fluctuations that destroy superconductivity. We took up the quasi-two-dimensional version [36, 39]. More precisely, at FIAN (the P.N. Lebedev Physical Institute of the USSR Academy of Sciences) myself and a group of theorists turned to the high-temperature superconductivity problem in the broad sense, considering all issues and possibilities known to us. The fruits of this labor were represented in the monograph [40]; even its English version (1982) appeared 4–5 years before the experimental realization of high-temperature superconductors [41, 42] in 1986–1987. If the consideration of different models and possibilities is omitted, the most significant quantitative finding of our work, which is primarily due to D.A. Kirzhnits, is the crystal stability condition. The point is that the main objection against the possibility of developing a high-temperature superconductor was the anxiety that the crystal lattice would be unstable for the metal parameter values required to obtain a high-temperature superconductor, i.e., for a material with $T_c > T_{b, N_2} = 77.4$ K.[7] When the problem is formulated in terms of the longitudinal material permittivity $\varepsilon(\omega, q)$, where ω is the frequency and q is the wave vector (we restrict our consideration to an isotropic body here), the production of electron pairs necessitates, roughly speaking, that the interelectron interaction $V = e^2/\varepsilon(0, q)r$ should be negative, i.e., should correspond to attraction. However, this corresponds to the requirement that $\varepsilon(0, q) < 0$. Meanwhile, on the grounds of some considerations it was believed that the lattice would be stable when

$$\varepsilon(0, q) > 0. \tag{1.10}$$

---

[6] Unfortunately, the outstanding theoretical physicist D.A. Kirzhnits died prematurely in 1998.

[7] I do not know whether there exists the commonly accepted definition of what superconductor can be regarded as a high-temperature one. In my opinion, HTSC takes place when $T_c > 77.4$ K, i.e., is higher than the boiling temperature of nitrogen at atmospheric pressure.

True, on closer examination (see [1, 40]) it was found that superconductivity is also possible under the condition (1.10), but the T_c values would turn out to be moderate, even below the estimate (1.9). In [40] and references therein it was found that the correct stability condition for $q \neq 0$ is of the form:

$$\frac{1}{\varepsilon(0, q)} \leq 1, \qquad (1.11)$$

i.e., is fulfilled when either of two inequalities:

$$\varepsilon(0, q) > 1, \quad \varepsilon(0, q) < 0 \qquad (1.12)$$

takes place. In other words, any negative values of $\varepsilon(0, q)$ are admissible from the standpoint of stability and there are no limitations on T_c. To be more precise, up to now we do not know of such limitations. The following conclusion was drawn from our work, which is contained in Chap. 1 in [40] (written by me):

'On the basis of general theoretical considerations, we believe at present that the most reasonable estimate is $T_c \lesssim 300$ K; this estimate being, of course, for materials and systems under more or less normal conditions (equilibrium or quasi-equilibrium metallic systems in the absence of pressure or under relatively low pressures, etc.). In this case, if we exclude from consideration metallic hydrogen and, perhaps, organic metals, as well as semimetals in states near the region of electronic phase transitions, then it is suggested that we should use the exciton mechanism of attraction between the conduction electrons.

In this scheme, the most promising materials – from the point of view of the possibility of raising T_c – are, apparently, layered compounds and dielectric–metal–dielectric sandwiches. However, the state of the theory, let alone the experiment, is still far from being such as to allow us to regard other possible directions as being closed, in particular, the use of filamentary compounds. Furthermore, for the present state of the problem of high-temperature superconductivity, the most sound and fruitful approach will be one that is not preconceived, in which attempts are made to move forward in the most diverse directions.

The investigation of the problem of high-temperature superconductivity is entering into the second decade of its history (if we are talking about the conscious search for materials with $T_c \gtrsim 90$ K with the use of the exciton and other mechanisms). Supposedly, there begins at the same time a new phase of these investigations, which is characterized not only by greater scope and diversity, but also by a significantly deeper understanding of the problems that arise. There is still no guarantee whatsoever that the efforts being made will lead to significant success, but a number of new superconducting materials have already been produced and are being investigated. Therefore, it is, in any case, difficult to doubt that further investigations of the problem of high-temperature superconductivity will yield many interesting results for physics and technology, even if materials that remain superconducting at liquid nitrogen (or even room) temperatures will not be produced. Besides, as has been emphasized, this ultimate aim does not seem to us to have been discredited in any way. As may be inferred, the next decade will be crucial for the problem of high-temperature superconductivity.'

This was written in 1976. Time passed, but the multiple attempts to find a reliable and reproducible way of creating a high-temperature superconductor have been unsuccessful. As a result, after the flash of activity came a slackening which gave cause for me to characterize the situation in a popular paper [43] published in 1984 as follows:

'It somehow happened that research into high-temperature superconductivity became unfashionable (there is good reason to speak of fashion in this context since fashion sometimes plays a significant part in research work and in the scientific community). It is hard to achieve anything by making admonitions. Typically it is some obvious success (or reports of success, even if erroneous) that can radically and rapidly reverse attitudes. When they smell success, the former doubters, and even dedicated critics, are capable of turning coat and becoming ardent supporters of the new work. But this subject belongs to the psychology and sociology of science and technology.

In short, the search for high-temperature superconductivity can readily lead to unexpected results and discoveries, especially since the predictions of the existing theory are rather vague.'

I did not expect, of course, that this 'prediction' would come true in two years [41, 42]. It came true not only in the sense that high-temperature superconductors with $T_c > T_{b, N_2} = 77.4$ K were obtained in a scientific aspect, but also, so-to-say, in a social aspect: as I have mentioned above, a real boom began and a 'high-temperature superconductivity psychosis' started. One of the manifestations of the boom and psychosis was the almost total oblivion of everything that had been done before 1986, as if the discussion of the high-temperature superconductivity problem had not begun 22 years before [35, 36]. I have already dwelt on this subject here and in the papers [44, 45] and do not want to return to it here. I will only note that J. Bardeen, whom I have always respected, treated the high-temperature superconductivity problem with understanding both before and after 1986 (see [46]; this article was also published in [16]).

These former remarks in no way imply that our group or I pretend to a practical contribution of great importance to the development of high-temperature superconductivity. At the same time, I believe that Little's works and ours have played a significant role in the formulation of the problem and have drawn attention to it. The solution of the problem was obtained to a large measure accidentally. The proposal to employ layered compounds was reasonable and promising, but neither I nor, to my knowledge, anybody else proposed precisely the use of the cuprates. Other layered compounds investigated do not belong to high-temperature superconductors. The following fact serves to illustrate the accidental, to a certain measure, character of dis-

covery of high-temperature superconductivity. As far back as 1979, one of
the institutes in Moscow produced and investigated [47] a $La_{1.8}Sr_{0.2}CuO_4$ ceramic, which was close to that investigated by Bednorz and Muller [41], with
$T_c \simeq 36$ K [48]. However, the authors of [47] measured the resistance of their
samples at temperatures not lower than the liquid-nitrogen temperature, and
therefore did not discover their superconductivity. From this one may draw a
trivial conclusion that all newly produced materials should be 'tested' for superconductivity. Also evident is another conclusion, namely, that even nowadays it is possible to make a major discovery and next year be awarded a
Nobel Prize for it without gigantic facilities and the work of a large group.
This should be a source of inspiration, particularly for young people.

The present situation in solid-state theory does not allow us either to
calculate the value of T_c or to calculate the value of other superconductor
parameters, with the possible exception of a metallic hydrogen yet to be produced. Moreover, for more than fifteen years the mechanism of superconductivity in cuprates has remained obscure. I should remark that, despite the fact
that I counted on the excitonic mechanism in high-temperature superconductivity research, the role of this mechanism in the known high-temperature
superconductors is still completely unclear. In this case, in high-temperature
superconductors (in cuprates) with $T_c < 170$ K (the highest-known value
$T_c \simeq 165$ K was attained back in 1994 in the $HgBa_2Ca_2Cu_3O_{8+x}$ cuprate
under high pressure), as I see it, the electron-phonon mechanism of pairing
may prove to be the dominant one. This possibility has previously been underestimated (in particular, by me), since the estimate (1.9) has served as a
guide. However, it is valid only for a weak coupling (1.7). For a strong coupling (i.e., when $\lambda_{eff} \gtrsim 1$), formula (1.6) is no longer applicable, but even from
this formula it is clear that T_c increases with λ_{eff}. The generalization of the
BCS theory [18] to the strong-coupling case [49] enables us to investigate the
corresponding possibilities. Their analysis (see particularly [50] and references
therein and in [1]) suggests that the electron-phonon mechanism in cuprates
may well ensure superconductivity with $T_c \lesssim 200$ K owing to the high θ_D
and λ_{eff} values. At the same time, the electron-phonon interaction alone is
supposedly insufficient in the context of the so-called d pairing and maybe
other special features of superconductivity in cuprates. However, the role of
other possibilities (such as spin interactions, excitonic interaction) is unclear.
Of course, it would be out of place to discuss this vital topical problem here. I
only want, on the one hand, to emphasize that the long-standing disregard of
electron-phonon interaction in cuprates has always seemed, and seems now,
unjustified to me (see [51]). On the other hand, the likelihood of attaining,
on the basis of the electron-phonon mechanism, the values $T_c \sim 300$ K, and
this is room-temperature superconductivity (RTSC), appears to be small[2*],
as with the use of the spin mechanism. At the same time, the excitonic mechanism, as far as I know, does not provoke objections for $T_c \sim 300$ K, either.
That is why I pin my hopes on precisely this mechanism for the attainment

of room-temperature superconductivity. However, all this is no more than an intuitive judgment.

The creation of high-temperature superconductivity had been my dream for 22 years, even with no guarantee that the goal was at all attainable and, in particular, attainable in the foreseeable future. In my view, obtaining room-temperature superconductivity now occupies the same place.

1.4 Thermoelectric Phenomena in the Superconducting State

The first attempt to observe thermoelectric phenomena, and, specifically, thermoelectric current or thermal electromotive force in a nonuniformly heated circuit of two superconductors, to my knowledge, was made by Meissner [52] in 1927. He arrived at the conclusion that the thermoelectric effect is completely absent in superconductors. When I took an interest in this problem in 1943, this viewpoint was generally accepted (see, for instance, [53] and especially the first and later editions of [25]). However, I encountered this statement more recently as well. Meanwhile, this conclusion is erroneous, which was pointed out in my work [11] which was published as far back as 1944.

The point is that the superconducting state can carry, apart from a superconducting current j_s, a normal current j_n as well. This normal current is carried by 'normal electrons', i.e., electron- or hole-type quasi-particles present in the metal in both the normal and superconducting states. In the superconducting state, the density of such normal quasi-particles depends strongly on the temperature and, generally, tends to zero as $T \to 0$. These notions, which are sometimes referred to as the two-liquid model, can be traced back to [54]. An isotropic non-superconductor or, more precisely, an isotropic metal residing in a normal state, can carry only the current with a density:

$$j = \sigma\left(E - \frac{\nabla\mu}{e}\right) + b\nabla T, \qquad (1.13)$$

where μ is the chemical potential of the electrons and E is the electric field. In the superconducting state, for a normal current we have (for more details, see [55]):

$$j_n = \sigma_n\left(E - \frac{\nabla\mu}{e}\right) + b_n\nabla T. \qquad (1.14)$$

At the same time, the superconducting current density j_s in the London theory [12] approximation, to which we restrict ourselves here (naturally, this is precisely the approximation used in [11]), obeys the equations

$$\mathrm{rot}\,(\Lambda j_s) = -\frac{1}{c}H, \qquad (1.15)$$

$$\frac{\partial(\Lambda \boldsymbol{j}_{\mathrm{s}})}{\partial t} = \boldsymbol{E} - \frac{\nabla \mu}{e}, \tag{1.16}$$

where $\Lambda = m/(e^2 n_{\mathrm{s}})$ is somewhat a constant, with n_{s} being the 'supercon-ducting electron' density (so that $\boldsymbol{j}_{\mathrm{s}} = e n_{\mathrm{s}} \boldsymbol{v}_{\mathrm{s}}$, where $\boldsymbol{v}_{\mathrm{s}}$ is the velocity); in this scheme, the field penetration depth is

$$\delta_{\mathrm{L}} = \sqrt{\frac{\Lambda c^2}{4\pi}} = \sqrt{\frac{mc^2}{4\pi e^2 n_{\mathrm{s}}}}.$$

Notice that this is a simplification, for different chemical potentials μ_{n} and μ_{s} should in fact be introduced in (1.14) and (1.16), respectively, for the normal and superconducting electrons. In addition, yet another term (generally, not large) proportional to ∇j_{s}^2 (see [55]) figures in (1.16). When the superconduc-tor is nonuniform, the parameter Λ depends on the coordinates.

As is clear from (1.16), in the stationary case, in the superconductor:

$$\boldsymbol{E} - \frac{\nabla \mu}{e} = 0 \tag{1.17}$$

and, in view of (1.14),

$$\boldsymbol{j}_{\mathrm{n}} = b_{\mathrm{n}}(T)\nabla T. \tag{1.18}$$

Therefore, the thermoelectric current $\boldsymbol{j}_{\mathrm{n}}$ in no way vanishes in the supercon-ducting state. However, this current in not directly observable in the simplest case, because it is compensated for by the superconducting current $\boldsymbol{j}_{\mathrm{s}}$. Let us consider a uniform superconducting rod, with one end of the rod residing at a temperature T_2 and the other at a temperature $T_1 < T_2$ (Fig. 1.4). Then, in the normal state (i.e., when $T_1 > T_{\mathrm{c}}$), since there is no closed circuit, from (1.13) we have (see Fig. 1.4a):

$$\boldsymbol{j} = 0, \quad \boldsymbol{E} - \frac{\nabla \mu}{e} = -\frac{b}{\sigma}\nabla T. \tag{1.19}$$

In the superconducting state (for $T_2 < T_{\mathrm{c}}$):

$$\boldsymbol{j} = \boldsymbol{j}_{\mathrm{s}} + \boldsymbol{j}_{\mathrm{n}} = 0, \quad \boldsymbol{j}_{\mathrm{s}} = -\boldsymbol{j}_{\mathrm{n}} = -b_{\mathrm{n}}\nabla T, \quad \boldsymbol{H} = 0, \quad \boldsymbol{E} - \frac{\nabla \mu}{e} = 0. \tag{1.20}$$

It is true, that near the rod ends, where $\boldsymbol{j}_{\mathrm{s}}$ transforms to $\boldsymbol{j}_{\mathrm{n}}$ or vice versa, uncompensated charges (the charge imbalance effect) emerge, and, therefore, the field $\boldsymbol{E}$ is not equal to $\nabla \mu/e$; in what follows I ignore this feature.

An important point is that the thermoelectric current $\boldsymbol{j}_{\mathrm{n}}$ exists in the uni-form case in the superconducting state (Fig. 1.4b), but the field $H = 0$. When the superconductor is nonuniform or anisotropic, the currents $\boldsymbol{j}_{\mathrm{s}}$ and $\boldsymbol{j}_{\mathrm{n}}$ do not in general compensate each other completely, and an observable thermo-electric magnetic field emerges, which was noted in [11]. In days of old (60

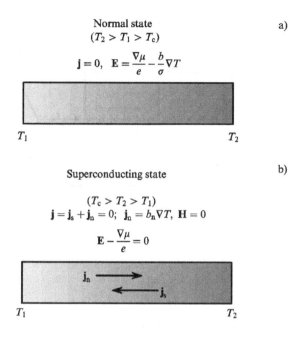

Fig. 1.4.

years ago!), as noted above, the case of alloys was considered to be unsavory and it was even unclear whether the London equation could be applied to alloys. That is why I restricted myself to a brief consideration of a bimetallic plate (say, of two different superconductors fused or soldered together: this juncture is the alloy) in the presence of a temperature gradient (see also §16 in [26] and [55]). In this case, because the parameter Λ depends on the coordinates (evidently, the Λ parameter is different for different metals), along the junction line there emerges an uncompensated current j, and hence the magnetic field H, which is perpendicular to the plate and the junction line (Fig. 1.5). Considered in greater detail in [11] and [26] was the case of an anisotropic superconductor. To this end, the London equations were generalized in a rather trivial way by replacing the scalar Λ with the tensor Λ_{ik} (for isotropic and cubic metals, $\Lambda_{ik} = \Lambda\delta_{ik}$). When the temperature gradient ∇T in a plate-shaped noncubic superconducting crystal is not directed along the symmetry axis, there emerges a current j flowing around the plate and a magnetic field H_T transverse to the plate and proportional to $(\nabla T)^2$. In principle, this field is not difficult to observe with modern techniques. Curiously enough, this is an interesting effect, which in addition makes it possible to measure the thermoelectric coefficient $b_n(T)$ or, more precisely, the components of its generalized tensor $b_{n,ik}(T)$. More than 30 years ago I managed to convince W. Fairbank to stage the corresponding experiment, and its results remain, as

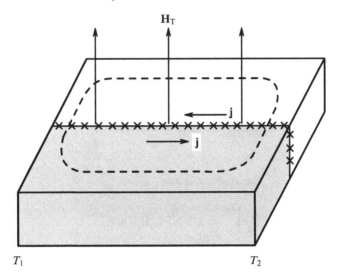

Fig. 1.5.

far as I know, the only ones on this subject [56]. Unfortunately, this work did not make things clear [55,57]. I am amazed by the fact that nobody has taken an interest in this question even after the fabrication of strongly anisotropic high-temperature superconductors. Evidently, such is the force of fashion in science, too.

It is true that a certain interest was attracted precisely by the isotropic superconductors, in essence, as applied to a more or less conventional thermoelectric current (Fig. 1.6a), for this circuit is equivalent to the 'circuit' of Fig. 1.6b. For this circuit it is easy to show [58,59] (the derivation is also given in [55]) that the magnetic flux $\Phi = \int \boldsymbol{H}\,\mathrm{d}\boldsymbol{S}$ through the opening is

$$\Phi = k\Phi_0 + \Phi_\mathrm{T}, \quad \Phi_\mathrm{T} = \frac{4\pi}{c}\int_{T_1}^{T_2}(b_{\mathrm{n,\,II}}\delta_\mathrm{II}^2 - b_{\mathrm{n,\,I}}\delta_\mathrm{I}^2)\,\mathrm{d}T,$$

$$\Phi_0 = \frac{hc}{2e} = 2 \times 10^{-7}\ \mathrm{G\ cm}^2, \quad k = 0, 1, 2, 3\ldots. \tag{1.21}$$

Here, the indices I and II refer to the superconducting metals I and II, δ_I and δ_II are the field penetration depths for these metals, $b_{\mathrm{n,\,I}}$ and $b_{\mathrm{n,\,II}}$ are the corresponding coefficients $b_\mathrm{n}(T)$ in (1.18), and Φ_0 is the so-called flux quantum. The configuration in Fig. 1.6b is essentially equivalent to the bimetallic plate in Fig. 1.5 with $k = 0$, i.e., without an opening. Unfortunately, I did not recognize this at the time (i.e., in [11, 26]).

If we assume for simplicity that

$$(b_\mathrm{n}\delta^2)_\mathrm{II} \gg (b_\mathrm{n}\delta^2)_\mathrm{I} \quad \text{and} \quad \delta_\mathrm{II}^2 = \delta_\mathrm{II}^2(0)(1 - T/T_{\mathrm{c,\,II}})^{-1},$$

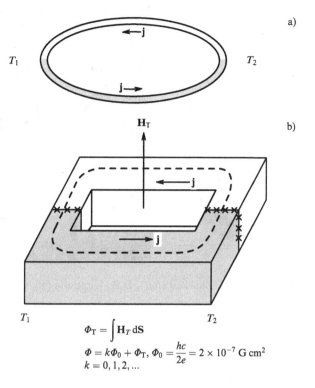

$$\Phi_T = \int \mathbf{H}_T \, d\mathbf{S}$$
$$\Phi = k\Phi_0 + \Phi_T, \ \Phi_0 = \frac{hc}{2e} = 2 \times 10^{-7} \ \text{G cm}^2$$
$$k = 0, 1, 2, \ldots$$

Fig. 1.6.

from expression (1.21) we obtain ($T_{c,\,II} = T_c$):

$$\Phi_T = \frac{4\pi}{c} \, b_{n,\,II}(T_c) \delta_{II}^2(0) T_c \ln\left(\frac{T_c - T_1}{T_c - T_2}\right). \tag{1.22}$$

If we substitute the known values $b_n(T_c)$ and $\delta(0)$ for $\ln\left(T_c - T_1\right)/(T_c - T_2) \sim 1$ in (1.22) we arrive at an estimate $\Phi_T \sim 10^{-2}\Phi_0$. This flux is easy to measure, which was done in several papers (see [1,55] and references therein). However, the flux Φ_T observed in some more complex configurations of the superconducting circuit was found to be orders of magnitude higher than the flux given by (1.21) and (1.22) and to possess a different temperature dependence [60]. The reason for this result has not been elucidated, and different assumptions have been made on that score [61,62]; see also other references in [1].

It is also pertinent to note that (1.21) and the ensuing (1.22) are obtained under the assumption that the equality $j = j_s + j_n = 0$ is fulfilled throughout the circuit depth (the current flows only near the surface). Meanwhile, as T_c is approached, the field penetration depth δ increases; as $T \to T_c$, the depth $\delta \to \infty$ and the current density j_n tends to the thermoelectric current density in the normal state, i.e., for $T > T_c$. In these conditions, a more detailed

Fig. 1.7. G.F. Zharkov (*left*), D.A. Kirzhnits (*right*)

analysis is required to include the charge imbalance effect. This interesting question has not been investigated (for more details, see [1]).

The aforesaid description is not the whole story. Even in the simplest case of a uniform superconductor, the existence of a temperature gradient (see Fig. 1.4b) affects the thermal conduction: since $j_n \neq 0$, there is bound to be an additional (convective) heat flux $q_c = -\varkappa_c \nabla T$ similar to that occurring in a superfluid liquid. This was noted even in [11] and was, in fact, the initial idea in this work.

The total heat flux in the superconducting state $q = -\varkappa \nabla T$, $\varkappa = \varkappa_{ph} + \varkappa_e + \varkappa_c$, where $\varkappa_{ph}$ is the thermal conductivity coefficient related to the lattice (phonons), $\varkappa_e$ is the electron contribution in the absence of convection (circulation), i.e., subject to the condition $j_n = 0$, and, as already noted, $\varkappa_c$ is the contribution of circulation. As is generally known, the thermal conductivity coefficient in the normal state is, by definition, measured for $j = 0$, and it is valid to say that $\varkappa_c = 0$ (see [8]). When estimating the $\varkappa_c$ coefficient, I, like others, got tangled up, and now I will restrict myself to a reference to paper [1] and a remark that in ordinary (not high-temperature) superconductors supposedly $\varkappa_c \ll \varkappa_e$. The role of $\varkappa_c$ in high-temperature superconductors is unclear to me. Most important of all, it is not clear how to extract $\varkappa_c$, even if it were possible to determine separately $\varkappa_{ph}$ and $\varkappa_{e,tot} = \varkappa_e + \varkappa_c$ (the total thermal conductivity coefficient $\varkappa$ is measured directly; on the separation of $\varkappa_{ph}$ from $\varkappa_{e,tot}$, see [1]).

---

[8] It is another matter that, for instance, a semiconductor subjected to the condition $j = 0$ in the presence of electron and hole conduction can simultaneously carry electron j_e and hole $j_h = -j_e$ currents; we ignore these possibilities.

Fig. 1.8. E.G. Maksimov

We have no way of dwelling on the thermoelectric effects in the superconducting state. My aim is to draw attention to this range of questions, which came under the scrutiny of science back in 1927 (see [52] as well as [25]) and under my scrutiny in 1944 [11], but which remains largely unclear to date[3*]. This is so in spite of a multitude of papers concerned with superconductivity.

1.5 Superfluidity Research and the Ψ-Theory of Superfluidity

Superconductivity is, if you please, the superfluidity of a charged liquid or, equivalently, superfluidity is the superconductivity of a noncharged liquid. It is, therefore, natural that the investigations of both effects have been interrelated. My first work in this area [8], concerned with light scattering in Helium II, was already mentioned in Sect. 1.2. By the way, there is good reason to revert to this question in light of modern understanding of the fluctuations near the λ point. Several other papers were dealt with in [1]; here, I will consider only the Ψ-theory of superfluidity, albeit with one exception. Namely, I

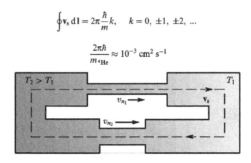

$$\oint \mathbf{v_s}\, d\mathbf{l} = 2\pi \frac{\hbar}{m} k, \quad k = 0, \pm 1, \pm 2, \dots$$

$$\frac{2\pi\hbar}{m_{^4\mathrm{He}}} \approx 10^{-3}\ \mathrm{cm^2\ s^{-1}}$$

Fig. 1.9.

would like to mention a proposal made jointly with A.A. Sobyanin[9] and partly with G.F. Zharkov [63, 64], and then mention the study of the feasibility of observing the thermomechanical circulation effect in a superfluid liquid.

In an annular vessel filled with a superfluid liquid (specifically, the case in point was Helium II), which has two different 'bottlenecks' (for instance, narrow capillaries), under a temperature gradient there is bound to emerge a circulation – a superfluid flow engulfing the entire vessel (Fig. 1.9). By the way, we made the inference about the existence of this effect [63] on the basis of analogy with the thermoelectric effect in a superconducting circuit. As to the inference about the existence of thermoelectric current in a superconducting circuit, I made it [11] at the time on the strength of an analogy with the behavior of Helium II under a temperature gradient. The above mentioned thermocirculation effect in Helium II has been observed [65] and discussed [64], and, in my view, interesting possibilities were pointed out for future research [64]. However, nobody, as far as I know, has taken an interest in this question during the past 20 years.

After the development of the Ψ-theory of superconductivity [3], the transfer of something similar to the superfluidity case appeared to be rather obvious. At the same time, even before (see, for instance [9]) I was concerned about the behavior of Helium II near the λ point, and the question of the boundary condition for superfluid component velocity $\mathbf{v_s}$ was obscure. By the way, L.D. Landau, the originator of the theory of phase transitions and superfluidity, was for some reason never concerned with this range of questions, as far as I know. In the Landau theory of superfluidity [7], the velocity $\mathbf{v_s}$ along the wall (unlike the normal component velocity $\mathbf{v_n}$) does not vanish at the wall: there is some kind of discontinuity. But in this case however, it seemed to me, that this discontinuity was bound to be related to some surface energy σ_s [66]. However, dedicated experiments [67] showed that the σ_s energy

---

[9] The talented theoretical physicist and public figure Aleksandr Sobyanin prematurely died at the age of 54 in 1997.

Fig. 1.10. L.P. Pitaevskii (*left*), A.A. Sobyanin (*right*)

is nonexistent or, in any case, is many orders of magnitude lower than the expected energy [66]. I saw a way out in the assumption that the superfluid component density at the wall $\rho_s(0)$ is zero. Then, the superfluid component flux $\boldsymbol{j}_s = \rho_s \boldsymbol{v}_s$ at the wall vanishes despite the fact that $\boldsymbol{v}_s$ has a discontinuity at the wall. In the Ψ-theory of superfluidity, evidently:

$$\rho_s = m|\Psi|^2, \tag{1.23}$$

where it may be assumed that $m = m_{\mathrm{He}}$ is the mass of a Helium atom (we imply the superfluidity of Helium II) and, in view of the former, the boundary condition at the wall is

$$\Psi(0) = 0, \tag{1.24}$$

instead of the condition (1.3) for superconductors. At this stage, as far as I remember, it turned out that L.P. Pitaevskii had independently taken up the Ψ-theory of superfluidity and, naturally, we combined efforts. As a result, the work [68] emerged; I speak of the Ψ-theory of superfluidity constructed in that work as 'initial' because I consider it below the 'generalized' Ψ-theory of superfluidity elaborated together with A.A. Sobyanin [69,70] (see also several other references in [1]).

The initial Ψ-theory of superfluidity [68] is quite similar to the Ψ-theory of superconductivity [3], of course, with the use of the boundary condition (1.24) and in the absence of the electric charge. In this case, the scalar complex function $\Psi = |\Psi| \exp(\mathrm{i}\varphi)$ obeys the equation

$$-\frac{\hbar^2}{2m}\Delta\Psi + \alpha(T)\Psi + \beta_\lambda|\Psi|^2\Psi = 0 \tag{1.25}$$

and

$$\boldsymbol{j}_{\mathrm{s}} = \rho_{\mathrm{s}}\boldsymbol{v}_{\mathrm{s}} = -\frac{\mathrm{i}\hbar}{2}(\Psi^{*}\nabla\Psi - \Psi\nabla\Psi^{*}) = \hbar|\Psi|^{2}\nabla\varphi, \qquad (1.26)$$

i.e., $\boldsymbol{v}_{\mathrm{s}} = (\hbar/m)\nabla\varphi$, with $m = m_{\mathrm{He}}$, irrespective of how Ψ is normalized (see [1,68]).

Furthermore, the correlation length ξ denoted as l in [68] is (T_{λ} is the temperature of the λ point):

$$\xi(T) = \frac{\hbar}{\sqrt{2m|\alpha|}} = \xi(0)\tau^{-1/2}, \quad \tau = \frac{T_{\lambda} - T}{T_{\lambda}}. \qquad (1.27)$$

The estimate of [68], based on experimental data, for ^{4}He, i.e., for Helium II, leads to a value $\xi(0) \sim 3 \times 10^{-8}$ cm. At the same time, the Ψ-theory is applicable only when the macroscopic Ψ function varies only slightly over atomic-scale distances. Hence, there follows the condition $\xi(T) \gg a \sim 3 \times 10^{-8}$ cm (here a is the average interatomic distance in liquid Helium). The Ψ-theory can therefore be adequate only near the λ point (for $\tau \ll 1$), say, for $(T_{\lambda} - T) < (0.1$–$0.2)$ K. A similar condition also takes place in the case of the Ψ-theory of superconductivity, which is also appropriate, generally speaking, only near T_{c}. It is of prime importance that the Landau theory of phase transitions, which is a mean-field theory, for superconductors (i.e., the Ψ-theory of superconductivity) is also correct in the immediate vicinity of T_{c}. This is due to the relatively large value of $\xi(0)$ in superconductors (the length $\xi(0)$ is on the order of the dimension of the Cooper pairs, i.e., in ordinary superconductors is on the order of, say, 10^{-5} cm). The point is that the temperature range near T_{c} (or T_{λ}), in which fluctuations are already large and the mean field approximation is inappropriate, is proportional to $[\xi(0)]^{-6}$ (see [1] and references therein, particularly [34]). In Helium II, the fluctuations near T_{λ} are relatively strong due to the smallness of $\xi(0)$, and the Ψ-theory [68] can be used only for $(T_{\lambda} - T) \gg 10^{-3}$ K [1]. Meanwhile, the temperature range significantly closer to T_{λ} is of special interest. That the mean-field theory is inapplicable in the region of the λ transition in ^{4}He is testified too by the very existence of the λ singularity in the temperature dependence of the heat capacity. This circumstance might not, at least on the face of it, be related to the temperature dependence of the density $\rho_{\mathrm{s}}(T)$, which was proportional to $|\Psi|^{2}$ [see (1.23)]. That is why in 1957, when the work [68] was carried out, we did not see the drawbacks to our theory right away. However, this became clear somewhat later, when it was found out that in Helium II to a good approximation:

$$\rho_{\mathrm{s}}(\tau) = \rho_{\mathrm{s}0}\tau^{\zeta}, \quad \zeta = \frac{2}{3}. \qquad (1.28)$$

In the mean-field theory,

$$\zeta = 1. \qquad (1.29)$$

In the experiment, by the way, the index ζ is not exactly equal to $2/3$ but is very close to it. For instance, according to [71], $\zeta = 0.6705 \pm 0.0006$.

Therefore, the initial Ψ-theory of superfluidity [68] is poorly applicable to liquid ^{4}He in a quantitative sense. At the same time, several results based on it were obtained in [68], which were also of significance for Helium II in a qualitative sense. The case in point is the density distribution $\rho_s(z)$ near the solid wall and in films with a thickness d in relation to this thickness. Also solved were the problems of velocity v_s circulation about a vortex line (filament) at the axis of which $\Psi = 0$, of the energy of this filament, and of the surface energy at the interface between Helium II and the solid wall. No less significant is the fact that liquid ^{4}He is not the only existing superfluid liquid. Such a liquid is also encountered in the case of ^{3}He–^{4}He solutions, liquid ^{3}He, neutron stars, and maybe in other cases. In all these cases, however, the Ψ function may prove to be no longer scalar but, on the other hand, the length $\xi(0)$ is relatively large (in liquid ^{3}He, for instance, $\xi(0) \sim 10^{-5}$ cm), and the fluctuation region is rather small. Finally, the theory of [68] had played, so far as I can judge, a significant role in the construction and elaboration of the Gross–Pitaevskii theory, which is widely used in the investigation of Bose–Einstein condensation (see [72]).

Liquid ^{4}He, i.e., Helium II, has always occupied and still occupies the leading position in the physics of superfluidity, both historically and regarding the scale of investigations. The Landau theory [7], which describes its behavior, is primarily macroscopic or, if you like, quasi-macroscopic. But it does not provide answers to several questions, particularly near the λ point. At the same time, a microtheory of the BCS type for superconductors does not exist for Helium II. On the other hand, Helium II, near the λ point, is interesting from various viewpoints, in particular, in the investigation of two-liquid hydrodynamics near the λ point, in the modeling of some cosmological situations [73], etc. It is likely that the initial Ψ-theory of superfluidity [68, 74] can be used to some extent for the solution of these problems, though with the above significant limitation arising from the inapplicability of the mean field approximation, i.e., from the neglect of fluctuations. The generalized Ψ-theory of superfluidity [69, 70] was intended to eliminate these drawbacks. It is based on some semiempirical generalization of the Landau theory of phase transitions (see, for instance, [75]). In the Landau theory of phase transitions and, in particular, in the Ψ-theory of superconductivity, i.e., when the Ψ function is selected as the order parameter, the free energy density of the ordered phase near the transition point T_λ is written in the form:

$$F_{II} = F_I + \alpha|\Psi|^2 + \frac{\beta}{2}|\Psi|^4 + \frac{\gamma}{6}|\Psi|^6; \tag{1.30}$$

away from the tricritical point, it being safe to assume that

$$\alpha = \alpha'_\lambda(T - T_\lambda) = -a_0\tau, \quad \beta = \beta_\lambda, \quad \gamma = 0, \quad \tau = \frac{T_\lambda - T}{T_\lambda}. \tag{1.31}$$

In the generalized theory

$$F_{\mathrm{II}} = F_{\mathrm{I}} - a_0\tau|\tau|^{1/3}|\Psi|^2 + \frac{b_0}{2}\,\tau^{2/3}|\Psi|^4 + \frac{g_0}{3}\,|\Psi|^6. \qquad (1.32)$$

When selecting (1.32), for small $|\Psi|^2$ in the equilibrium $|\Psi_0|^2 = -\alpha/\beta = (a_0/b_0)\tau^{2/3}$, i.e., there occurs a temperature dependence which agrees with the observed one [see (1.28)]. Evidently, (1.32) is selected for precisely the attainment of this goal.

The generalized Ψ-theory of superfluidity [69, 70] formally differs from the initial theory [68, 74] just by the replacement of expressions (1.30), (1.31) with (1.32). Several expressions and inferences were derived on this basis. For instance, for a thin film of Helium II of thickness d, the λ-transition temperature is:

$$T_\lambda(d) = T_\lambda - 2.53 \times 10^{-11}\left(\frac{3+M}{M}\right)d^{-3/2}\ \mathrm{K}, \qquad (1.33)$$

where $T_\lambda = T_\lambda(\infty)$ is the λ-transition temperature in massive Helium (as is well known, $T_\lambda = 2.17$ K) and M is the parameter of the theory proportional to the g_0 coefficient in (1.32). When $M < 1$, the λ transition is of the second kind (in comparison with the experiment, only a crude estimate was obtained for Helium II: $M = 0.5\pm0.3$). By the way, if we consider a cylindrical capillary of diameter d instead of a plane film, the coefficient 2.53 in (1.33) should be replaced with 4.76. Quite a number of other expressions were also derived [69, 70, 76].

Unfortunately, the generalized Ψ-theory of superfluidity has not come to the attention of either experimentalists or theorists. True, some pessimistic judgments regarding it were expressed in the literature (they were mentioned in [1]). A.A. Sobyanin and I also abandoned the superfluidity research during the period of rapid changes in the USSR and Russia that set in after 1985–1988. Only in [1] did I review our work.

Undeniably the generalized Ψ-theory of superfluidity is not a lofty ab initio theory. At the same time, its simplicity (at least in comparison with other known methods) suggests that the Ψ-theory of superfluidity (initial as well as generalized) can still yield much in the study of superfluidity. In any case, the opposite opinion is not substantiated at all. This section of the lecture has been written precisely with the aim of attracting the attention of physicists engaged in the corresponding areas to the Ψ-theory of superfluidity. It may well be, in my view, that the lack of attention is a delusion. It is also conceivable, on the contrary, that I am in error myself.

1.6 The 'Physical Minimum':
What Problems of Physics and Astrophysics
Seem Now to be Especially Important and Interesting
in the Beginning of the 21st Century?

I have encountered the viewpoint that my work in the area of superconductivity and superfluidity is a matter of the remote past. There is no question that the work of Ginzburg and Landau [3] performed back in 1950 stands out. But on the whole, as is clear from the my previous remarks and particularly from [1], I have been occupied with this field of physics since 1943 until the present time. In this case, it seems to me, several questions and problems have also been posed which have not been solved and which deserve attention. Of course, presently the most urgent problems in the area of superconductivity are the elucidation of the mechanism and several features of high-temperature superconductivity, and the creation of room-temperature superconductivity. More precisely, what is wanted in the latter case is to elucidate the potentialities and formation conditions of room-temperature superconductors. I am keenly aware that I will not be able to accomplish anything in the last two directions. I would like only to witness as many new findings as possible.

That is why, in recent years, I have been placing progressively stronger emphasis, as far as physics is concerned, on some educational program, which I conventionally call the 'physical minimum'. As far as I know, many young scientists attend Nobel Lectures, and therefore I decided to enlarge on this 'physical minimum'. I believe that this will be of greater interest to young people than to hear what was going on before they were born.

Physics has developed rapidly and fruitfully, especially in the past century. Its face changed radically even within a human life span. I myself was already 16 when the neutron and positron were discovered in 1932. And what would modern physics be without neutrons and positrons? As a result of so rapid a development, physics and its adjacent realms (for instance, astronomy) have enormously expanded, both as regards their basic contents and their bodies of information. In the recent past it was possible to be guided by the requirement "to know something about everything and to know everything about something" (say, in physics), but now, it seems to me, this is no longer possible. At the same time, I am startled and dispirited when young physicists (and sometimes not so young ones) restrict themselves to the knowledge in 'their' areas and are not informed, or only informed in a general way, about the state of physics as a whole and its 'hottest' areas. This situation cannot be justified by alleging an absence of a pivot (keystone) in contemporary physics or its boundlessness. Quite the contrary. Physics does (maybe still does) have its pivot, which is represented by fundamental concepts and laws formulated in theoretical physics. It is possible, on the basis of theoretical physics studied during one's student days, to understand all modern physics or, more precisely, to understand how matters stand everywhere in physics and be aware of the situation. Every physicist (naturally, this equally applies to other specialities, but I restrict myself to physicists for definitiveness) should simultaneously know, apart from theoretical physics, a wealth of facts from different branches of physics and be familiar with the newest notable accomplishments.

At the same time, we in Russia like to quote a certain Koz'ma Prutkov, a fictitious character, who said pompously, in particular, that "there is no way of comprehending the incomprehensible". So, one has to choose something. And so I took this path: I have made a 'list' of the top problems of the day. Any such 'list' is admittedly subjective. It is also clear that the 'list' should vary with time. It is also clear that subjects not included in the 'list' can in no way be regarded as unimportant or uninteresting. It is simply that many of them presently seem less pressing to me (or to the authors of other similar lists). But again, "one cannot comprehend the incomprehensible". Those who know interesting subjects beyond the 'list' have no reason to be offended and should only supplement or change the 'list'. I only suggest some enumeration of the questions of which, in my view, every physicist should have at least a superficial idea. Supposedly less trivial is the statement that this is not as difficult as it might seem at first glance. The time to be spent for this purpose is, I believe, no longer than the time a good student spends preparing for an examination, say, on electrodynamics. Acquaintance with all subjects included in this 'list' is what I call the 'physical minimum'. Of course, this 'minimum' is the echo of the 'theoretical minimum' proposed by Landau in the 1930s. It is significant that there are many excellent textbooks on electrodynamics (or other subjects in the university curriculum), among which the corresponding volume of the *Course of Theoretical Physics* by L.D. Landau and E.M. Lifshitz ranks, in my view, as the highest. But a beginner needs help to get acquainted with the 'physical minimum'. Working out this 'list', as well as commenting on it, has served and hopefully continues to serve precisely this purpose. In 1995, in the Russian edition of the book [16], I managed to work out a rather detailed commentary. But in the English translation [16] some information was already out of date, which I failed to compensate for in full measure. Inserted at the beginning of the book [2] is an article also concerned with the 'physical minimum'. Several additional remarks were introduced in the English translation of this book, which will hopefully be published soon. On the whole, should the proposal be taken advantage of and elaborated, the 'physical minimum' will meet with support, and new books on this subject should appear. Unfortunately, I cannot set myself to this task.

In the context of this lecture it only remains for me to recall the well-known saying that the proof of the pudding is in the eating and give the above-mentioned 'list' for the beginning of the 21st century:

1. Controlled nuclear fusion.
2. High-temperature and room-temperature superconductivity (HTSC and RTSC).

3. Metallic hydrogen, and other exotic substances.
4. Two-dimensional electron liquid (the anomalous Hall effect and other effects).
5. Some questions of solid-state physics (heterostructures in semiconductors, quantum wells and dots, metal–dielectric transitions, charge and spin density waves, mesoscopics).
6. Second-order and related phase transitions. Some examples of such transitions. Cooling (in particular, laser cooling) to superlow temperatures. Bose–Einstein condensation in gases.
7. Surface physics. Clusters.
8. Liquid crystals. Ferroelectrics. Ferrotoroics.
9. Fullerenes. Nanotubes.
10. The behavior of matter in superstrong magnetic fields.
11. Nonlinear physics. Turbulence. Solitons. Chaos. Strange attractors.
12. X-ray lasers, gamma-ray lasers, superhigh-power lasers.
13. Superheavy elements. Exotic nuclei.
14. Mass spectrum. Quarks and gluons. Quantum chromodynamics. Quark-gluon plasma.
15. The unified theory of weak and electromagnetic interactions. $W^{\pm}$ and Z^0-bosons. Leptons.
16. The standard model. Grand unification. Superunification. Proton decay. Neutrino mass. Magnetic monopoles.
17. Fundamental length. Particle interaction at high and superhigh energies. Colliders.
18. Nonconservation of CP-invariance.
19. Nonlinear phenomena in a vacuum and in superstrong magnetic fields. Phase transitions in a vacuum.
20. Strings. M-theory.
21. The experimental verification of the general theory of relativity.
22. Gravitational waves and their detection.
23. The cosmological problem. Inflation. Λ-term and 'quintessence'. The relationship between cosmology and high-energy physics.
24. Neutron stars and pulsars. Supernova stars.
25. Black holes. Cosmic strings (?).
26. Quasars and galactic nuclei. The formation of galaxies.
27. The problem of dark matter (hidden mass) and its detection.
28. The origin of superhigh-energy cosmic rays.
29. Gamma-bursts. Hypernovae.
30. Neutrino physics and astronomy. Neutrino oscillations.

The singling out of 30 particular problems (more precisely, items in the 'list') is of course absolutely conditional. Moreover, some of them might be subdivided. In my first 'list' published in 1971 there were 17 problems [77]. Subsequently their number would grow (for some more details, see [2]). It would supposedly be good to add some new subjects to the 'list', for instance, those concerning quantum computers and advances in optics. But I cannot do this with adequate comprehension.

Any 'list' is undoubtedly not a dogma; something can be discarded and something added, depending on the preferences of lecturers and authors of corresponding papers. More interesting is the question of the evolution of the 'list' with time, as it reflects the development of physics. In the 'list' of 1970–1971 [77] quarks were given only three lines in the enumeration of the attempts to explain the mass spectrum. This did not testify to my perspicacity. However, at that time (in 1970) quarks were only five or six years old (I mean the age of the corresponding hypothesis), and the fate of the concept of the quark was indeed vague. Now the situation is of course quite different. True, the heaviest t-quark was discovered only in 1994 (its mass, according to the data of 1999, is $m_t = 176 \pm 6$ GeV). The list [77] naturally contains no fullerenes, which were discovered in 1985, and no gamma-bursts (the first report of their discovery was published in 1973). High-temperature superconductors were synthesized in 1986–1987, but in the list [77] this problem was nonetheless considered rather thoroughly for it had been discussed since 1964 (this was discussed in greater detail in the previous sections of the lecture). Generally, much has been done in physics over the past 30 or 35 years, but, I believe, not very much essentially new has appeared. In any case, the 'lists' in [16, 77], as well as that presented above, characterize to a certain extent the development and the state of physical and astronomical problems from 1970–1971 to the present day.

It should be added that three 'great problems' of modern physics are also to be included in the 'physics-minimum', included in the sense that they should be singled out in some way and specially discussed, and the development of these 'great problems' should be reviewed. This is discussed at some length in [2]. The 'great problems' are, first, the increase in entropy, time irreversibility, and the 'time arrow'. Second is the problem of interpretation of nonrelativistic quantum mechanics and the possibility of learning something new even in the field of its applicability (I personally doubt this possibility but believe that one's eyes should remain open). And third is the question of live-to-liveless reduction, i.e., the feasibility of explaining the origin of life and thought on the basis of physics alone. On the face of it, how could it be otherwise? But until the questions are elucidated, one cannot be quite sure of anything. I think that the problem of the origin of life will unreservedly be solved only after 'life in a test-tube' is created. Until then, this will be an open question.

One more concluding remark. In the past century, and even nowadays, one could encounter the opinion that in physics nearly everything had been done. There allegedly are only dim 'cloudlets' in the sky or theory, which will soon be eliminated to give rise to the 'theory of everything'. I consider these views as some kind of blindness. The entire history of physics, as well as the state of present-day physics and, in particular, astrophysics, testifies to the opposite. In my view we are facing a boundless sea of unresolved problems.

It only remains for me to envy the younger members of the audience, who will witness a great deal of new, important, and interesting things.

1.7 Notes

1*. V.L. Ginzburg's 2003 Nobel Lecture in Physics was first published in English in the annual publication by the Nobel Foundation *Les Prix Nobel / The Nobel Prizes 2003*; in Russian it was first published in the Journal "Uspekhi Fizicheskikh Nauk" **174** 1240 (2004).

2*. See also V.V. Schmidt, *The Physics of Superconductors: Introduction to Fundamentals and Applications* (Eds. P. Müller, A.V. Ustinov) (New York: Springer, 1997); W. Buckel, R. Kleiner, *Superconductivity* (Wiley–VCH, 2004).

3*. Research in the area of thermoelectric effects in superconductors has intensified recently (see, for instance, [78–81]).

References

1. V.L. Ginzburg, Usp. Fiz. Nauk **167**, 429, 1997; **168**, 363, 1998 [Phys. Usp. **40**, 407, 1997; **41**, 307, 1998]. (It is also Chap. 2 in this book.)
2. V.L. Ginzburg, *O Nauke, o Sebe i o Drugikh (About Science, Myself, and Others)*, Fizmatlit, Moscow, 2003[10] [Translated into English, IOP Publ., Bristol, 2005, Chap. 7].
3. V.L. Ginzburg, L.D. Landau, Zh. Eksp. Teor. Fiz. **20**, 1064, 1950; This paper was published in English in the volume: L.D. Landau, *Collected Papers*, p. 546, Pergamon Press, Oxford, 1965. (See Chap. 4 below.)
4. A.A. Abrikosov, Zh. Eksp. Teor. Fiz. **32**, 1442, 1957 [Sov. Phys. JETP **5**, 1174, 1957].
5. E.M. Lifshitz, L.P. Pitaevskii, *Statisticheskaya Fizika (Statistical Physics)* Pt. 2, Teoriya Kondensirovannogo Sostoyaniya (Theory of Condensed State), Nauka, Moscow, 1978, 1999 [Translated into English, Pergamon Press, Oxford, 1980].
6. M. Tinkham, *Introduction to Superconductivity*, 2nd edn., McGraw Hill, New York, 1996.
7. L.D. Landau, Zh. Eksp. Teor. Fiz. **11**, 592, 1941; J. Phys. USSR **5**, 71, 1941.
8. V.L. Ginzburg, Zh. Eksp. Teor. Fiz. **13**, 243, 1943; J. Phys. USSR **7**, 305, 1943.
9. V.L. Ginzburg, Zh. Eksp. Teor. Fiz. **14**, 134, 1944.
10. J. Bardeen, in *Kältephysik* (Handbuch der Physik, Bd. 15, Hrsg. S. von Flügge, p. 274, Springer, Berlin, 1956 [Translated into Russian: J. Bardeen, in *Fizika Nizkikh Temperatur*, ed. by A.I. Shal'nikov, p. 679, IL, Moscow, 1959].

---

[10] Chapter 7 published in this collection is the somewhat edited article cited in [1] and Chap. 2 in this book.

11. V.L. Ginzburg, Zh. Eksp. Teor. Fiz. **14**, 177, 1944; J. Phys. USSR **8**, 148, 1944.
12. F. London, H. London, Proc. R. Soc. London Ser. A **149**, 71, 1935; Physica **2**, 341, 1935.
13. J.R. Waldram, *Superconductivity of Metals and Cuprates*, Institute of Physics Publ., Bristol, 1996.
14. V.L. Ginzburg, Zh. Eksp. Teor. Fiz. **16**, 87, 1946; J. Phys. USSR **9**, 305, 1945.
15. V.L. Ginzburg, Zh. Eksp. Teor. Fiz. **15**, 739, 1945; J. Phys. USSR **10**, 107, 1946.
16. V.L. Ginzburg, *The Physics of a Lifetime. Reflections on the Problems and Personalities of 20th Century Physics*, Springer-Verlag, Berlin, 2001.[11]
17. V.L. Ginzburg, Zh. Eksp. Teor. Fiz. **29**, 748, 1955 [Sov. Phys. JETP **2**, 589, 1956].
18. J. Bardeen, L.N. Cooper, and J.R. Schrieffer, Phys. Rev. **108**, 1175, 1957.
19. V.L. Ginzburg, Usp. Fiz. Nauk **48**, 25, 1952; Fortschr. Phys. **1**, 101, 1953.
20. R.A. (Jr.) Ogg, Phys. Rev. **69**, 243; **70**, 93, 1946.
21. M.R. Schafroth, Phys. Rev. **96**, 1149, 1954; **100**, 463, 1955.
22. L.N. Cooper, Phys. Rev. **104**, 1189, 1956.
23. L.P. Gor'kov, Zh. Eksp. Teor. Fiz. **36**, 1918; **37**, 1407, 1959 [Sov. Phys. JETP **9**, 1364, 1959; **10**, 998, 1960].
24. V.L. Ginzburg, Zh. Eksp. Teor. Fiz. **31**, 541, 1956 [Sov. Phys. JETP **4**, 594, 1957].
25. D. Shoenberg, *Superconductivity*, 3rd edn., Cambridge Univ. Press, Cambridge, 1965 [Translated into Russian, IL, Moscow, 1955].
26. V.L. Ginzburg, *Sverkhprovodimost' (Superconductivity)*, Izd. AN SSSR, Moscow–Leningrad, 1946.
27. W. Buckel, *Supraleitung*, Physik-Verlag, Weinheim, 1972 [Translated into English: W. Buckel, *Superconductivity: Fundamentals and Applications*, VCH, Weinheim, 1991; Translated into Russian: Mir, Moscow, 1975].
28. V.L. Ginzburg, Zh. Eksp. Teor. Fiz. **23**, 236, 1952.
29. V.L. Ginzburg, Zh. Eksp. Teor. Fiz. **34**, 113, 1958 [Sov. Phys. JETP **7**, 78, 1958].
30. V.L. Ginzburg, Zh. Eksp. Teor. Fiz. **42**, 299, 1962 [Sov. Phys. JETP **15**, 207, 1962].
31. V.L. Ginzburg, Zh. Eksp. Teor. Fiz. **36**, 1930, 1959 [Sov. Phys. JETP **9**, 1372, 1959].
32. V.L. Ginzburg, Zh. Eksp. Teor. Fiz. **31**, 202, 1956 [Sov. Phys. JETP **4**, 153, 1957].
33. G.F. Zharkov, Zh. Eksp. Teor. Fiz. **34**, 412, 1958; **37**, 1784, 1959 [Sov. Phys. JETP **7**, 278, 1958; **10**, 1257, 1959].
34. V.L. Ginzburg, Fiz. Tverd. Tela **2**, 2031, 1960 [Sov. Phys. Solid State **2**, 1824, 1961].
35. W.A. Little, Phys. Rev. **134**, A1416, 1964.
36. V.L. Ginzburg, Phys. Lett. **13**, 101, 1964; Zh. Eksp. Teor. Fiz. **47**, 2318, 1964 [Sov. Phys. JETP **20**, 1549, 1965].
37. V.L. Ginzburg, D.A. Kirzhnits, Zh. Eksp. Teor. Fiz. **46**, 397, 1964 [Sov. Phys. JETP **19**, 269, 1964].

---

[11] This book is, for the most part, a translation of my book *O Fizike i Astrofizike*, Byuro Kvantum, Moscow, 1995.

38. V.L. Ginzburg, Phys. Scripta **T27**, 76, 1989.
39. V.L. Ginzburg, D.A. Kirzhnits, Dokl. Akad. Nauk SSSR **176**, 553, 1967 [Sov. Phys. Dokl. **12**, 880, 1968].
40. V.L. Ginzburg, D.A. Kirzhnits (Eds.), *Problema Vysokotemperaturnoi Sverkhprovodimosti (The Problem of High-Temperature Superconductivity)*, Nauka, Moscow, 1977 [Translated into English: V.L. Ginzburg, D.A. Kirzhnits (Eds.), *High-Temperature Superconductivity*, Consultants Bureau, New York, 1982].
41. J.G. Bednorz, K.A. Muller, Z. Phys. B **64**, 189, 1986.
42. M.K. Wu, J.R. Ashburn, C.J. Torng, P.H. Hor, R.L. Meng, L. Gao, Z.J. Huang, Y.Q. Wang, C.W. Chu, Phys. Rev. Lett. **58**, 908, 1987.
43. V.L. Ginzburg, *Energiya* **9**, 2, 1984.
44. V.L. Ginzburg, Prog. Low Temp. Phys. **12**, 1, 1989.
45. V.L. Ginzburg, in *From High-Temperature Superconductivity to Microminiature Refrigeration*, ed. by B. Cabrera, H. Gutfreund, V. Kresin, Plenum Press, New York, 1996.
46. V.L. Ginzburg, J. Supercond. **4**, 327, 1986.
47. I.S. Shaplygin, B.G. Kakhan, V.B. Lazarev, Zh. Neorg. Khim. **24**, 1476, 1979.
48. R.J. Cava et al., Phys. Rev. Lett. **58**, 408, 1987.
49. G.M. Eliashberg, Zh. Eksp. Teor. Fiz. **38**, 966; **39**, 1437, 1960 [Sov. Phys. JETP **11**, 696, 1960; **12**, 1000, 1961].
50. E.G. Maksimov, Usp. Fiz. Nauk **170**, 1033, 2000 [Phys. Usp. **43**, 965, 2000].
51. V.L. Ginzburg, E.G. Maksimov, Sverkhprovodimost': Fiz., Khim., Tekh. **5**, 1543, 1992 [Superconductivity: Phys., Chem., Technol. **5**, 1505, 1992].
52. W.Z. Meissner, Ges. Kälteindustr. **34**, 197, 1927.
53. E.F. Burton, G.H. Smith, and J.O. Wilhelm, *Phenomena at the Temperature of Liquid Helium* (American Chemical Society: Monograph Ser., No. 83), Reinhold, New York, 1940.
54. C.J. Gorter, H. Casimir, Phys. Z. **35**, 963, 1934.
55. V.L. Ginzburg, G.F. Zharkov, Usp. Fiz. Nauk **125**, 19, 1978 [Sov. Phys. Usp. **21**, 381, 1978].
56. P.M. Selzer, W.M. Fairbank, Phys. Lett. A **48**, 279, 1974.
57. V.L. Ginzburg, G.F. Zharkov, Pis'ma Zh. Eksp. Teor. Fiz. **20**, 658, 1974 [JETP Lett. **20**, 302, 1974].
58. Yu.M. Gal'perin, V.L. Gurevich, V.N. Kozub, Zh. Eksp. Teor. Fiz. **66**, 1387, 1974 [Sov. Phys. JETP **39**, 680, 1974].
59. J.C. Garland, D.J. Van Harlingen, Phys. Lett. A **47**, 423, 1974.
60. D.J. Van Harlingen, Physica B+C **109–110**, 1710, 1982.
61. R.M. Arutyunyan, V.L. Ginzburg, G.F. Zharkov, Zh. Eksp. Teor. Fiz. **111**, 2175, 1997 [JETP **84**, 1186, 1997]; Usp. Fiz. Nauk **167**, 457, 1997 [Phys. Usp. **40**, 435, 1997].
62. Y.M. Galperin et al., Phys. Rev. B **65**, 064531, 2002.
63. V.L. Ginzburg, G.F. Zharkov, A.A. Sobyanin, Pis'ma Zh. Eksp. Teor. Fiz. **20**, 223, 1974 [JETP Lett. **20**, 97, 1974]; V.L. Ginzburg, A.A. Sobyanin, G.F. Zharkov, Phys. Lett. A **87**, 107, 1981.
64. V.L. Ginzburg, A.A. Sobyanin, Zh. Eksp. Teor. Fiz. **85**, 1606, 1983 [Sov. Phys. JETP **58**, 934, 1983].
65. G.A. Gamtsemlidze, M.I. Mirzoeva, Zh. Eksp. Teor. Fiz. **79**, 921, 1980; **84**, 1725, 1983 [Sov. Phys. JETP **52**, 468, 1980; **57**, 1006, 1983].
66. V.L. Ginzburg, Zh. Eksp. Teor. Fiz. **29**, 254, 1955 [Sov. Phys. JETP **2**, 170, 1956].

67. G.A. Gamtsemlidze, Zh. Eksp. Teor. Fiz. **34**, 1434, 1958 [Sov. Phys. JETP **7**, 992, 1958].
68. V.L. Ginzburg, L.P. Pitaevskii, Zh. Eksp. Teor. Fiz. **34**, 1240, 1958 [Sov. Phys. JETP **7**, 858, 1958].
69. V.L. Ginzburg, A.A. Sobyanin, Usp. Fiz. Nauk **120**, 153, 1976 [Sov. Phys. Usp. **19**, 773, 1976]; J. Low Temp. Phys. **49**, 507, 1982.
70. V.L. Ginzburg, A.A. Sobyanin, Usp. Fiz. Nauk **154**, 545, 1988 [Sov. Phys. Usp. **31**, 289, 1988]; Jpn. J. Appl. Phys. **26** (Suppl. 26–3), 1785, 1987.
71. L.S. Golder, N. Mulders, G.J. Ahlers, Low. Temp. Phys. **93**, 131, 1992.
72. L. Pitaevskii, S. Stringari, *Bose-Einstein Condensation* (Intern. Series of Monographs on Physics, Vol. 116), Clarendon, Oxford, 2003.
73. W.H. Zurek, Nature **382**, 296, 1996.
74. L.P. Pitaevskii, Zh. Eksp. Teor. Fiz. **35**, 408, 1958 [Sov. Phys. JETP **8**, 282, 1959].
75. Yu.G. Mamaladze, Zh. Eksp. Teor. Fiz. **52**, 729, 1967 [Sov. Phys. JETP **25**, 479, 1967]; Phys. Lett. A **27**, 322, 1968.
76. V.L. Ginzburg, A.A. Sobyanin, in *Superconductivity, Superdiamagnetism, Superfluidity*, ed. by V.L. Ginzburg, p. 242, MIR Publ., Moscow, 1987.
77. V.L. Ginzburg, Usp. Fiz. Nauk **103**, 87, 1971 [Sov. Phys. Usp. **14**, 21, 1971].
78. G.F. Zharkov, Usp. Fiz. Nauk **174**, 1012, 2004 [Phys. Usp. **47**, 944, 2004].
79. T. Löfwander, M. Fogelström, Phys. Rev. B **70**, 0245515, 2004.
80. L. Koláček, P. Lipavsky, Phys. Rev. B **71**, 092503, 2005.
81. A. Maniv et al., Phys. Rev. Lett. **94**, 247005, 2005.

2

Superconductivity and Superfluidity
(What was Done and What was Not Done)[1*]

2.1 Introduction. Early Works

I, the author of the present paper, am 80 years old (this paper was written in 1997) and I cannot hope to obtain new, important scientific results. At the same time, I feel I need to summarize my work of over 50 years. I do not mean now my work in general (I have been engaged in solving quite a variety of physical and astrophysical problems, see [1], p. 312 and also Chap. 6 below) but my activity in the field of superconductivity and superfluidity. In general, it is not traditional to write such papers. In my opinion, however, this comes from a certain prejudice. In any case, I decided to try and write such a paper, something like a scientific autobiography, but devoted only to two related problems – superconductivity and superfluidity. I may say that it is not associated with some priority or any other claims: it is only a desire to continue my work, though in an unusual form. I leave it to the reader to judge whether this attempt has been pertinent and successful.

I began working, i.e., obtaining some results in physics in 1938–1939, when I graduated from the physics faculty of Moscow State University. Before the Second World War, i.e., until mid-1941, I was engaged in classical and quantum electrodynamics, as well as the theory of higher-spin particles. We somehow felt that war would break out and were scared of it, but were unprepared, and lived with the hope that the danger would pass. I am not going to generalize, but this atmosphere reigned in the Department of Theoretical Physics of FIAN (the P.N. Lebedev Physical Institute of the USSR Academy of Sciences). When the danger did not pass by, we began looking, while waiting for the call-up or some other changes in our lives, for an application of our abilities which might be of use for defense. I, for one, was engaged in problems of radio-wave propagation in the ionosphere (see [1, 2]). But these and similar subjects remained, at least in my case, far from finding an application in defense. Therefore, I went on working in various fields under these or other influences.

The most important such influence, not to mention the continuation of research in the field of the relativistic theory of spin particles, was exerted by L.D. Landau. In 1939, after a year's confinement in prison, Landau started working on the theory of the superfluidity of Helium II.[1] I was present, if I am not mistaken, in 1940, at Landau's talk devoted to this theory (the corresponding paper, [4], was submitted for publication in 15 May, 1941). At the end of the paper [4], he also considered superconductivity interpreted as the superfluidity of an electron liquid in a metal. I do not know whether an assertion of the kind had ever been expressed before but it is hardly probable. (Some hint, in this respect, was made in [5].) The point is that superfluidity in the proper sense of the word was discovered only in 1938 independently and simultaneously by P.L. Kapitza [5] and G.F. Allen and A.D. Misener [6].

We mean here a frictionless flow through capillaries and gaps. As to the anomalous behavior of liquid Helium (^{4}He) below the λ-point, i.e., at a temperature $T < T_\lambda = 2.17\,\mathrm{K}$, the study of this issue began, in effect, in 1911. Precisely in the year when superconductivity was discovered [7] (for more details, see [8,9]; paper [7] is also included in [9] as an appendix), Kamerlingh Onnes reported a Helium density maximum at T_λ [10,11,127]. It was only in 1928 that the existence of two phases – Helium I and Helium II – became obvious and, in 1932, a clear λ-shaped curve for the temperature dependence of the specific heat near the λ-point was obtained. The superhigh thermal conductivity of Helium II was discovered by W. Keezom and A. Keezom (see the references in [11,12]) in 1936 and, finally, superfluidity was revealed [5,6] in 1938. One can thus say that it took 27 years (from 1911 to 1938) to discover superfluidity [127]. Such a long process is in obvious contrast with the discovery of superconductivity, which was practically a one-stroke occurrence [7] (for details, see [8,9] and Chap. 6 in [2]). One can hardly doubt that the reason lies in the different methods. Superconductivity was discovered when the electrical resistance of a wire (or, more precisely, a capillary filled with mercury) was being measured. It is a much more difficult task to investigate the character of liquid flow (concretely, Helium II) through gaps and capillaries and, besides, one must hit upon the idea of carrying out such experiments.

At the same time, the origin of superfluidity remained obscure. Landau believed [4] that the responsibility rested with the spectrum of 'elementary excitations' in a liquid, while Bose statistics and Bose–Einstein condensation had nothing to do with it. F. London and L. Tisza [12], on the contrary, associated superfluidity with Bose–Einstein condensation. The validity of the latter opinion became obvious in 1949 after liquid ^{3}He with atoms obeying Fermi statistics, the properties differing radically from those of liquid ^{4}He, had been obtained. Theoretically, the same conclusion was drawn by Feynman (see [13]). But nothing could be derived from it in respect of superconductivity because

---

[1] As is well known, P.L. Kapitza's plea for Landau's discharge from prison was motivated by the very wish to have his assistance in the field of superfluidity theory (see [3]).

electrons obey Fermi statistics. As we know today, the solution of the problem (or rather the puzzle) lies in the fact that electrons in a superconductor form 'pairs' with zero spin. Such pairs can undergo Bose–Einstein condensation with which the transition to a superconducting state is associated. My fairly modest contribution to this subject consists in pointing out that, in a Bose gas of charged particles, the Meissner effect must be observed [14]. The idea of 'pairing' itself did not occur to me. To the best of my knowledge, R.A. Ogg was the first to suggest it in 1946 [15]. This viewpoint was supported and further developed by M.R. Schafroth [16]. However, the cause and mechanism of pairing remained absolutely vague, and it was only in 1956 that L.N. Cooper [17] pointed out a concrete mechanism of pairing in a Fermi gas with attracting particles. This was the basis on which J. Bardeen, L.N. Cooper, and J.R. Schrieffer (BCS) finally formulated the first consistent, though model-type microtheory of superconductivity [18] in 1957. It is curious that [18] contains no indications of Bose–Einstein condensation, while it is, in fact, the crucial point.

However, I am running many years ahead as far as my own work is concerned. Concretely, in 1943, I tried [19], on the basis of the Landau theory [4] of superfluidity, to construct a quasi-microscopic theory of superconductivity [19]. The paper postulated a spectrum of electrons (charged 'excitations') in a metal with a gap Δ. For such a spectrum, superconductivity (the superfluidity of a charged liquid) must be observed. The introduction of a gap provided the critical field with a dependence on temperature and penetration depth into a superconductor, which approximately corresponded to the actual one. A comparison between the theory and the experiment gave the value $\Delta/k_B T_c = 3.1$. As is well known, in BCS theory $2\Delta_0/k_B T_c = 3.52$ but the most important point is that $\Delta_0 \equiv \Delta(0)$ is the value of the gap at $T = 0$ and, with increasing temperature, the gap decreases to yield $\Delta(T_c) = 0$. In my paper, the gap Δ was assumed to be constant and a satisfactory agreement with the experiment is possibly explained by the inaccuracy of the experimental data employed. I do not think that a more detailed analysis of this question is pertinent because model [19] is of no more than historical value now. Nonetheless, [19] did have some ideas that could have been of interest; for example, the occurrence of resonance phenomena for incident radiation at a frequency $\nu = \Delta/h$ was mentioned. In any case, the fact is that in his well-known review [20], published in 1956, Bardeen covered the results of paper [19] rather extensively. Notice that paper [19] also presented a survey of the macrotheory of superconductivity. It was followed by [21] considering gyromagnetic and electron inertia experiments with superconductors. Finally, in the same year, 1944, paper [22], devoted to thermoelectric phenomena in superconductors, was published.[2] This latter paper remains topical even now and we shall return to it in Sect. 2.5. The previously mentioned papers [19, 21, 22] were in-

---

[2] It should be noted that all three papers [19, 21, 22] were submitted for publication on the same date (23 November, 1943). I do not remember why this happened.

cluded in the monograph *Superconductivity* [24] written in 1944. Before taking up superconductivity, I analyzed [23], on the basis of the Landau theory, the problem of light scattering in Helium II. In what follows, I shall consider this and some other papers devoted to superfluidity (see Sect. 2.6).

2.2 The Ψ-Theory of Superconductivity (The Ginzburg–Landau Theory)

Within the first two decades after the discovery of superconductivity, its study went rather slowly compared to today's standards. This does not seem strange if we remember that liquid helium, which was first obtained in Leiden in 1908, became available elsewhere only after 15 years, i.e., in 1923. Without plunging into the history (see [8, 9, 11]; see also Chap. 6 in [2]), I shall restrict myself to the remark that the Meissner effect was only discovered [25] in 1933, i.e., 22 years after the discovery of superconductivity. Only after that did it become clear that a metal in normal and superconducting states can be treated as two phases of a substance in the thermodynamic sense of this notion. As a result, in 1934, there appeared [20, 26] the so-called two-fluid approach to superconductors and also the relation:

$$F_{n0}(T) - F_{s0}(T) = \frac{H_{cm}^2(T)}{8\pi},\tag{2.1}$$

where F_{n0} and F_{s0} are free-energy densities (in the absence of a field) in the normal and superconducting phase, respectively, and H_{cm} is the critical magnetic field destroying superconductivity. A differentiation of expression (2.1) with respect to T leads to expressions for the differences of entropy and specific heat.

According to the two-fluid picture, the total electric current density in a superconductor is

$$j = j_s + j_n,\tag{2.2}$$

where j_s and j_n are the densities of the superconducting and normal current.

The normal current in a superconductor does not, in fact, differ from the current in a normal metal and, in the local approximation, we have

$$j_n = \sigma_n(T)\,E,\tag{2.3}$$

where E is the electric field strength and σ_n is conductivity of the 'normal part' of the electron liquid; for simplicity, we henceforth take $j_n = 0$, unless otherwise specified.

In 1935, F. London and H. London proposed [27] for j_s the equations (now referred to as Londons' equations):

Most probably it was connected with some special conditions pertaining to the war.

$$\text{rot}\,(\Lambda\,\boldsymbol{j}_{\mathrm{s}}) = -\frac{1}{c}\,\boldsymbol{H} \tag{2.4}$$

$$\frac{\partial(\Lambda\,\boldsymbol{j}_{\mathrm{s}})}{\partial t} = \boldsymbol{E}, \tag{2.5}$$

where Λ is a constant and the magnetic field strength $\boldsymbol{H}$ here and later does not differ from the magnetic induction $\boldsymbol{B}$.

We arrive at such equations, for example, proceeding from the hydrodynamic equations for a conducting 'liquid' which consists of particles with charge e, mass m, and velocity $\boldsymbol{v}_{\mathrm{s}}(\boldsymbol{r}, t)$:

$$\begin{aligned}
\frac{\partial\boldsymbol{v}_{\mathrm{s}}}{\partial t} &= -(\boldsymbol{v}_{\mathrm{s}}\nabla)\boldsymbol{v}_{\mathrm{s}} + \frac{e}{m}\,\boldsymbol{E} + \frac{e}{mc}\,[\boldsymbol{v}_{\mathrm{s}}\boldsymbol{H}] \\
&= \frac{e}{m}\,\boldsymbol{E} - \nabla\frac{\boldsymbol{v}_{\mathrm{s}}^{2}}{2} + \left[\boldsymbol{v}_{\mathrm{s}}\left(\text{rot}\,\boldsymbol{v}_{\mathrm{s}} + \frac{e}{mc}\,\boldsymbol{H}\right)\right].
\end{aligned} \tag{2.6}$$

Such an equation corresponds to infinite (ideal) conductivity [28] and is not an obstruction to the presence of a constant magnetic field in a superconductor, which contradicts the existence of the Meissner effect. Therefore, the Londons imposed, so to say, an additional condition $\text{rot}\,\boldsymbol{v}_{\mathrm{s}} + e\boldsymbol{H}/mc = 0$, interpreted as the condition of a vortex-free motion for a charged liquid. If $\boldsymbol{j}_{\mathrm{s}}$ is written in the form $\boldsymbol{j}_{\mathrm{s}} = en_{\mathrm{s}}\boldsymbol{v}_{\mathrm{s}}$, where n_{s} is the charge concentration, the additional condition for $n_{\mathrm{s}} = \text{const}$ assumes precisely form (2.4) and

$$\Lambda = \frac{m}{e^{2}n_{\mathrm{s}}}. \tag{2.7}$$

Equation (2.6) transforms to (2.5) up to a small term proportional to $\nabla\boldsymbol{v}_{\mathrm{s}}^{2}$ (see Sect. 2.5). Within such an approach, the principal Londons' equation (2.4) is, of course, merely postulated. This condition is an effect of quantum nature and follows from the Ψ-theory of superconductivity [29] considered later and from the microtheory of superconductivity [18, 30] which, in turn, transforms near T_{c} to Ψ-theory [31]).

Londons' equation (2.4), along with the Maxwell equation

$$\text{rot}\,\boldsymbol{H} = \frac{4\pi}{c}\,\boldsymbol{j}_{\mathrm{s}} \tag{2.8}$$

at $\Lambda = \text{const}$ (we are obviously dealing with the quasi-stationary case), leads to the equations:

$$\Delta\boldsymbol{H} - \frac{1}{\delta^{2}}\,\boldsymbol{H} = 0, \quad \Delta\boldsymbol{j}_{\mathrm{s}} - \frac{1}{\delta^{2}}\,\boldsymbol{j}_{\mathrm{s}} = 0 \tag{2.9}$$

$$\delta^{2} = \frac{\Lambda c^{2}}{4\pi} = \frac{mc^{2}}{4\pi e^{2}n_{\mathrm{s}}}. \tag{2.10}$$

Equation (2.9) implies that the magnetic field $\boldsymbol{H}$ and the current density $\boldsymbol{j}_{\mathrm{s}}$ exponentially decay through the superconductor depth (for example, in the field parallel to and near a flat boundary, we have $H = H_0 \exp(-z/\delta)$, where z is the distance from the boundary), i.e., the Meissner effect arises. The Londons' equations still hold true but only in the case of a weak field:

$$H \ll H_c, \tag{2.11}$$

where H_c is the critical magnetic field destroying superconductivity (in the case of non-local coupling between the current and the field, Londons' equations do not hold either [20, 30] but we do not consider such cases here). We mean here type I superconductors. For type II superconductors, the Londons' theory has a wider limit of applicability, including the vortex phase for $H \ll H_{c2}$ at any temperature. But if the field is strong, i.e., comparable with H_c, Londons' theory may be invalid or otherwise insufficient. So, from Londons' theory, it follows that the critical magnetic field H_c, in which the superconductivity of a flat film of thickness $2d$ is destroyed (in the field parallel to it), is

$$H_c = \left(1 - \frac{\delta}{d} \mathrm{th}\frac{d}{\delta} \right)^{-1/2} H_{cm},$$

where H_{cm} is the critical field for a massive specimen (see [24, 32, 33] and references therein). This expression for H_c, however, contradicts experimental data. The situation can be improved by introducing different surface tensions σ_n and σ_s on the boundaries of the normal and the superconducting phases with a vacuum [32]. It turns out, however, that

$$\frac{\sigma_n - \sigma_s}{H_{cm}^2/8\pi} \sim \delta \sim 10^{-5} \ (\mathrm{cm}).$$

σ is a conductivity in (2.3) but I hope this will not lead to any misunderstanding.

At the same time, it might be expected that $(\sigma_n - \sigma_s) \sim (10^{-7}\text{--}10^{-8})$ $H_{cm}^2/8\pi$, i.e., is of the order of the volume energy $H_{cm}^2/8\pi$ multiplied by an atomic scale length. Moreover, in the theory based on Londons' equations, on the boundary between the normal and superconducting phases, the surface tension (surface energy) connected with the field and the current is $\sigma_{ns}^{(0)} = -\delta\, H_{cm}^2/8\pi$. Consequently, to obtain a positive surface tension $\sigma_{ns} = \sigma_{ns}^{(0)} + \sigma_{ns}^{(\prime)}$ observed for a stable boundary, it is necessary to introduce a certain surface energy $\sigma_{ns}^{(\prime)} > \delta H_{cm}^2/8\pi$ of non-magnetic origin. The introduction of such a comparatively high energy is totally ungrounded. On the contrary, one can think that a rational theory of superconductivity must automatically lead to the possibility of expressing the energy σ_{ns} in terms of parameters characterizing the superconductor.

Such a theory that generalized the Londons' theory, eliminated the indicated difficulties, and suggested some new conclusions, was the Ψ-theory [29] formulated in 1950.[3] In the same year, I wrote a review [33] devoted to the macro-theory of superconductivity, including the Ψ-theory.

In the absence of a magnetic field, the superconducting transition is a second-order transition. The general theory of such transitions always includes [34] a certain parameter (the order parameter) η which, when in equilibrium, differs from zero in the ordered phase and equals zero in the disordered phase. For example, in the case of ferroelectrics, the role of η is played by the spontaneous electric polarization $\boldsymbol{P}_s$ and, in the case of magnetics, by the spontaneous magnetization $\boldsymbol{M}_s$ (not long before the appearance of our paper [29]; both these cases were discussed in the review [35]). In superconductors, where the ordered phase is superconducting, for the order parameter we chose a complex function Ψ which plays the role of an 'effective wavefunction of superconducting electrons'. This function can be so normalized that $|\Psi|^2$ is the concentration n_s of 'superconducting electrons.'

The free energy density of a superconductor and the field was written in the form:

$$F_{sH} = F_{s0} + \frac{H^2}{8\pi} + \frac{1}{2m}\left| -i\hbar\nabla\Psi - \frac{e}{c}\boldsymbol{A}\Psi \right|^2 ,$$

$$F_{s0} = F_{n0} + \alpha|\Psi|^2 + \frac{\beta}{2}|\Psi|^4, \qquad (2.12)$$

where $\boldsymbol{A}$ is the vector potential of the field $\boldsymbol{H} = \mathrm{rot}\,\boldsymbol{A}$. Without the field, in the state of thermodynamic equilibrium $\partial F_{s0}/\partial|\Psi|^2 = 0$, $\partial^2 F_{s0}/\partial^2|\Psi|^2 > 0$ and we must have $|\Psi|^2 = 0$ for $T > T_c$ and $|\Psi|^2 > 0$ for $T < T_c$. This implies that $\alpha_c \equiv \alpha(T_c) = 0$ and $\beta_c \equiv \beta(T_c) > 0$, and $\alpha < 0$ for $T < T_c$. Within the validity limits of expansion (2.12) in $|\Psi|^2$, one can put $\alpha = \alpha'_c(T - T_c)$ and $\beta(T) = \beta_{T_c} \equiv \beta_c$. From this, at $T < T_c$ [see also (2.1)], we have:

$$|\Psi|^2 \equiv |\Psi_\infty|^2 = -\frac{\alpha}{\beta} = \frac{\alpha'_c(T_c - T)}{\beta_c},$$

$$F_{s0} = F_{n0} - \frac{\alpha^2}{2\beta} = F_{n0} - \frac{(\alpha'_c)^2(T_c - T)^2}{2\beta_c} = F_{n0} - \frac{H^2_{cm}}{8\pi}. \qquad (2.13)$$

In the presence of the field, the equation for Ψ is derived upon varying the free energy $\int F_{sH}dV$ with respect to Ψ^* and, obviously, has the form:

$$\frac{1}{2m}\left(-i\hbar\nabla - \frac{e}{c}\boldsymbol{A} \right)^2 \Psi + \alpha\Psi + \beta|\Psi|^2\Psi = 0. \qquad (2.14)$$

---

[3] This theory is usually called the Ginzburg–Landau theory. I try, however, to avoid this term, not out of false modesty but rather because in such cases the use of one's own name is not conventional in Russian. Furthermore, in its application to superfluidity (not superconductivity) the Ψ-theory was developed not with L.D. Landau but with L.P. Pitaevskii and A.A. Sobyanin (see Sect. 2.4). The article [29] is included in this book.

If, on the superconductor boundary, the variation $\delta\Psi^*$ is arbitrary, i.e., no additional condition is imposed on Ψ and no additional term corresponding to the surface energy is introduced in (2.12), then the condition of minimal free energy is the so-called natural boundary condition on the superconductor boundary:

$$\boldsymbol{n}\left(-i\hbar\nabla\Psi - \frac{e}{c}\boldsymbol{A}\Psi\right) = 0, \tag{2.15}$$

where $\boldsymbol{n}$ is the normal to the boundary (for more details, see [29] and Sect. 2.3). Condition (2.15) refers to the case of a boundary between a superconductor and a vacuum or a dielectric. As regards the equation for $\boldsymbol{A}$, under the condition $\operatorname{div}\boldsymbol{A} = 0$, and after variation of the integral $\int F_{sH}dV$ over $\boldsymbol{A}$, it becomes

$$\Delta\boldsymbol{A} = -\frac{4\pi}{c}\boldsymbol{j}_s, \quad \boldsymbol{j}_s = -\frac{ie\hbar}{2m}(\Psi^*\nabla\Psi - \Psi\nabla\Psi^*) - \frac{e^2}{mc}|\Psi|^2\boldsymbol{A}. \tag{2.16}$$

Here, of course, we assume that $\boldsymbol{j}_n = 0$, i.e., the total current is superconducting. An expression similar to (2.14) is, of course, also obtained for Ψ^* and, as expected, we have $\boldsymbol{j}_s\boldsymbol{n} = 0$ on the boundary [see (2.15)]. The solution of the problem of the distribution of the field, current, and function Ψ in a superconductor is reduced to the integration of the system of (2.14) and (2.16). An expression similar to (2.14) is, of course, also obtained for Ψ^* and, as expected, we have $\boldsymbol{j}_s\boldsymbol{n} = 0$ on the boundary [see (2.15)]. The solution of the problem of the distribution of the field, current, and function Ψ in a superconductor is reduced to the integration of the system of (2.14) and (2.16). Assuming $\Psi = \Psi_\infty = \text{const}$, the superconducting current density is $\boldsymbol{j}_s = -e^2|\Psi_\infty|^2\boldsymbol{A}/mc = -e^2 n_s\boldsymbol{A}/mc$ (with normalization $|\Psi_\infty|^2 = n_s$). Applying the operation rot to this expression, we obtain Londons' equation (2.4) [see also (2.7)]. Thus, the Ψ-theory generalizes the Londons' theory and passes over into it in the limiting case $\Psi = \Psi_\infty = \text{const}$.

Paper [29] is rather long (19 pages) and solves several problems to which we shall return in what follows. After that, I myself, sometimes with co-authors, devoted a number of papers to the development of the Ψ-theory of superconductivity. These papers are mentioned later. Moreover, this theory was further promoted and accounted for in a lot of papers and books (see, for example, [20,30,33,36–41,229,236–239]). I do not follow the corresponding literature now, the more so as (2.14) and its extensions are widely used outside superconductivity or only in applications to superconductors (see, for example, [42–44]). This equation is also being investigated by mathematicians whose works are incomprehensible to me (see, for example, [45]). The relativistic generalization of the equations of the Ψ-theory and some of the concepts associated with this theory also enjoy wide applications in quantum field theory (for example, spontaneous symmetry breaking, etc; see [46]).[4] In

---

[4] To confirm this, I would cite paper [46] (see p. 184; p. 480 in the English translation): 'It is easy to see that the Higgs model is fully analogical to the Ginzburg–

such a situation, it seems absolutely impossible to elucidate here the present-day state of the Ψ-theory or even focus in detail on the original paper [29] and my subsequent papers.

However, what I think is necessary is to tell the story of the appearance of paper [29] and to speak about the role of Landau and myself. Nobody else can do this because regretfully Lev Davidovich Landau passed away long ago (he stopped working in 1962 and died in 1968). At the same time, this is, of course, a very delicate question. That is why, when 20–25 years ago I was approached by the bibliographical magazine *Current Contents* with a request to elucidate the history of the appearance of [29], I refused. My refusal was motivated by the fact that my story might be interpreted as an attempt to exaggerate my role. And, in general, I had no desire to prove that I was indeed a full co-author and not a student or a postgraduate to whom Landau 'had set a task', whilst actually doing everything himself. Without such a premise it is difficult to explain why our paper has been frequently cited as Landau and Ginzburg, although it is known to have Ginzburg and Landau in the title. Of course, I have never made protestations concerning this point and, in general, consider it to be a trifle, but still I believe that such a citation with a wrong order of authors is incorrect. It would certainly still be incorrect even if my role had indeed been a secondary one. But I did not think so, and neither did Landau, and this fact was well known to his circle, and generally in the USSR. As to foreigners, they really did not know much about scientific research in the USSR at that time, for in 1950 the Cold War was at its height. As far back as 1947, the *USSR Journal of Physics*, which was a good journal, stopped being published and [29] appeared only in Russian. We could not go abroad at that time. Perhaps we sent a reprint to D. Shoenberg or he himself came across this article in *Zh. Eksp. Teor. Fiz* (JETP). In any event, Shoenberg translated the paper into English on his own initiative, and distributed it among some people and it then became available at least to some colleagues. Landau's name played, of course, a positive role and stimulated a lively interest in the paper.

One way or another, I decided to dwell on the history of the appearance of the work [29] because the present paper would be incomplete if I did not.

I regard the already mentioned paper [32] as being accomplished as far back as 1944 (it was submitted for publication on 21 December, 1944), as initial. From [32], it is quite clear that the London theory is invalid for the description of the behavior of superconductors in strong enough fields and, in particular, for the calculation of the critical field in the case of films. The introduction of the surface energies σ_n and σ_s was an artificial technique, and these quantities were absurdly large new constants whose values were not predicted by the the-

---

Landau theory and is its relativistic generalization. It turned out that this conclusion bears an important heuristic value by allowing to establish direct analogs between superconductivity theory and theories of elementary particles, including the Higgs model.' (See also [265].)

ory. The same applies to the surface energy σ_{ns} on the boundary between the normal and superconducting phases. It was also absolutely unclear how the critical current should be calculated in the case of small-sized superconductors. Therefore, it was necessary somehow to generalize the Londons' theory to overcome its limits. Unfortunately, advancement in this direction was slow. One of the possible explanations is that, like many theoretical physicists of my generation and the previous, I was simultaneously engaged in the solution of various problems and did not concentrate on anything definite (it can be seen, for instance, from the bibliographical index [47]). But there was gradual progress. So, on the basis of the conception of the Landau theory [4], I came to the conclusion [48] that electromagnetic processes in superconductors must be nonlinear and, incidentally, suggested a possible experiment for revealing such nonlinearity. The main point is that, in note [48], I made the following remark: 'The indication of a possible inadequacy of the classical description of superconducting currents consists in the fact that the zero energy of excitation in a superconductor is equal in order of magnitude to $\hbar^2 n/m\delta \sim 1\,\mathrm{erg\,cm^{-2}}$ (for $\delta \sim 10^{-5}\mathrm{cm}$ and $n \sim 10^{22}\mathrm{cm^{-3}}$) and is thus higher than the magnetic energy $\delta H^2/8\pi \sim 0.1\,\mathrm{erg\,cm^{-2}}$ (for $H \sim 500\,\mathrm{Oe.}$)' The feeling that the theory of superconductivity should take into account quantum effects was also reflected in [49], devoted for the most part to critical velocity in Helium II. At the same time, in that paper, I also tried to apply the theory of second-order phase transitions to the λ-transition in liquid helium.

It seems surprising, and unfortunately, it did not occur to me at that time to ask why Landau, the author of the theory of phase transitions [34] and the theory of superfluidity [4], had never posed the question of the order parameter η for liquid helium. In [49], I chose as such a parameter ρ_s, i.e., the density of the superfluid phase of Helium II. However, this choice raises doubts because the expansion of the free energy (thermodynamic potential) begins with the term $\alpha\rho_s$, whereas, in the general theory, the first term of the expansion has the form $\alpha\eta^2$. Hence, $\sqrt{\rho_s}$ is a more pertinent choice as the order parameter. But $\sqrt{\rho_s}$ is proportional to a certain wavefunction Ψ, so far as it is precisely the quantity $|\Psi|^2$ that is proportional to the particle concentration. Unfortunately, I do not remember exactly whether or not it was these arguments alone that prompted me to introduce the order parameter $\eta = \Psi$ and nothing is said about it in [49]. More important for me was the desire to explain the surface tension σ_{ns} by the gradient term $|\nabla\Psi|^2$. In quantum mechanics, this term has the form of kinetic energy $\hbar^2|\nabla\Psi|^2/2m$. It was precisely this idea that I suggested to Landau, probably in late 1949 (paper [29] was submitted on 20 April, 1950 but it had taken a great deal of time to prepare it). I was on good terms with Landau; I attended his seminars and often asked his advice on various problems. Landau supported my idea of introducing the 'effective wavefunction Ψ of superconducting electrons' as the order parameter, and so we were immediately led to the free energy (2.12). The thing I do not remember exactly (and certainly do not want to contrive) is whether I came to him with the ready expression:

$$\frac{1}{2m}\left| -\mathrm{i}\hbar\nabla\Psi - \frac{e}{c}\boldsymbol{A}\Psi \right|^2$$

or with an expression without the vector potential. The introduction of the latter is obvious by analogy with quantum mechanics, but perhaps this was made only during a conversation with Landau. I feel I should present my apologies to the reader for such reservations and uncertainty but since that time nearly 50 years have passed (!), no notes have remained, and I never thought that I would have to recall those remote days.

After the basic equations (2.12), (2.14), and (2.16) of the Ψ-theory were derived, one had to solve various problems on their basis and compare the theory with experiments. Naturally, it was myself who was mostly concerned with this but I regularly met with Landau to discuss the results. What has been said may produce the impression that my role in the creation of the Ψ-theory was even greater than that of Landau. But this is not so. One should not forget that the fundamental basis was the theory [34,50] of second-order phase transitions developed by Landau in 1937 which I had employed in a number of cases [35, 49] and applied to the theory of superconductivity in paper [29]. Moreover, I find it necessary to note that the important remark made in [29] concerning the meaning of the Ψ-function used as an order parameter was due to Landau himself. I shall cite the relevant passage from [29]:

"Our function $\Psi(\boldsymbol{r})$ may be thought of as immediately related to the density matrix $\rho(\boldsymbol{r}, \boldsymbol{r}') = \int \Psi^*(\boldsymbol{r}, \boldsymbol{r}'_i)\,\Psi(\boldsymbol{r}', \boldsymbol{r}'_i)\,\mathrm{d}\boldsymbol{r}'_i$, where $\Psi(\boldsymbol{r}, \boldsymbol{r}'_i)$ is the true Ψ-function of electrons in a metal which depends on the coordinates of all the electrons $\boldsymbol{r}_i$ ($i = 1, 2, \ldots, N$) and $\boldsymbol{r}'_i$ are the coordinates of all the electrons except a distinguished one (its coordinates are $\boldsymbol{r}$ and at another point $\boldsymbol{r}'$). One may think that for a non-superconducting body, where the long range order is absent, as $|\boldsymbol{r} - \boldsymbol{r}'| \to \infty$ we have $\rho \to 0$, while in the superconducting state $\rho(|\boldsymbol{r} - \boldsymbol{r}'| \to \infty) \to \rho_0 \neq 0$. In this case it is natural to assume the density matrix to be related to the introduced Ψ-function as $\rho(\boldsymbol{r}, \boldsymbol{r}') = \Psi^*(\boldsymbol{r})\Psi(\boldsymbol{r}')$."

Accordingly, the superconducting (or superfluid) phase is characterized by a certain long-range order which is absent in ordinary liquids (see also [30], Sect. 26; [51,52], [53], Sect. 9.7.). This result is usually ascribed to C.N. Yang [51] and is referred to as off diagonal long-range order (ODLRO) [53]. However, as we can see, Landau realized the possibility of the existence of this long-range order 12 years before Yang. I mentioned this fact in [54].

In (2.12) and subsequent expressions, the coefficients e and m appear. These designations were, of course, chosen by analogy with the quantum-mechanical expression for the Hamiltonian of a particle with charge e and mass m. Our Ψ-function is, however, not the wavefunction of electrons. The coefficient m can be taken arbitrarily [29] because the Ψ-function is not an observed quantity: an observed quantity is the penetration depth δ_0 of a weak magnetic field (see (2.12), (2.13), and (2.16)):

$$\delta_0^2 = \frac{mc^2\beta_{\mathrm{c}}}{4\pi e^2|\alpha|} = \frac{mc^2}{4\pi e^2|\Psi_\infty|^2}. \tag{2.17}$$

Since the Ψ-theory in a weak field (2.11) transforms to the London theory (though a number of problems cannot be stated in the London theory even in this case), the penetration depth δ_0 is frequently called the London penetration depth and is denoted by δ_L or λ_L. If we assume [29] e and m to correspond to a free electron ($e_0 = 4.8 \times 10^{-10}$CGS, $m_0 = 9.1 \times 10^{-28}$g), then $|\Psi_\infty|^2 = n_s$, where n_s is the 'superconducting electron' concentration thus defined. In fact, one can choose any arbitrary value of m [29, 37] which will only affect the normalization of the observed quantity $|\Psi_\infty|^2$. In the literature, $m = 2m_0$ occasionally occurs, which corresponds to the mass of a 'pair' of two electrons. As to the charge e in (2.12) and subsequent expressions, it is an observed quantity (see later). It seemed to me from the very beginning that one should regard the charge e in (2.12) as a certain 'effective charge' e_{eff} and take it as a free parameter. But Landau objected and, in paper [29], it is stated as a compromise that 'there is no reason to assume the charge e to be other than the electron charge'. Running ahead, I shall note that I still went on thinking of the question of the role of the charge $e \equiv e_{eff}$ as open and pointed out the possibility of clarifying the situation by comparing the theory with the experiment (see [14], p. 107). The point is that the essential parameter involved in the Ψ-theory is the quantity:

$$\varkappa = \frac{mc}{e\hbar} \sqrt{\frac{\beta_c}{2\pi}} = \frac{\sqrt{2}\,e}{\hbar c} H_{cm}\delta_0^2. \tag{2.18}$$

In [29], we set $e = e_0$ and could, therefore, determine $\varkappa$ from experimental data on H_{cm} and δ_0. At the same time, the parameter $\varkappa$ enters the expressions for the surface energy σ_{ns}, for the penetration depth in a strong field ($H \gtrsim H_{cm}$) and the expressions for superheating and supercooling limits. Using the approximate data of measurements available at the time, I came to the conclusion [55] (this paper was submitted for publication on 12 August, 1954) that the charge $e \equiv e_{eff}$ in (2.18) is two to three times greater than e_0. When I discussed this result with Landau, he put forward a serious objection to the possibility of introducing an effective charge (he had apparently had this argument in mind before, when we discussed paper [29] but did not then advance it). Specifically, the effective charge might depend on the composition of a substance, its temperature and pressure, and, therefore, might appear to be a function of coordinates. But, in that case, the gradient invariance of the theory would be broken, which is inadmissible. I could not find arguments against this remark and, with the consent of Landau, I included it in paper [55]. The explanation seems now to be quite simple. No, an effective charge e_{eff}, which might appear to be coordinate-dependent, should not have been introduced. But it might well be supposed that, say, $e_{eff} = 2e_0$. And this was exactly the case, but it became obvious only after the creation of BCS theory [18] in 1957, and after the appearance of the paper by L.P. Gorkov [31] who showed that the Ψ-theory near T_c follows from the BCS theory. More precisely, the Ψ-theory near T_c is certainly wider than the BCS theory in the sense that it is independent of some particular assumptions used in the BCS theory. But this is a different subject. The formation of pairs with charge $2e_0$ is a very general phenomenon, too. I have already emphasized that the idea of pairing and, what is important, the realistic character of such pairing, was far from trivial.

So, in the Ψ-theory, we have $e = 2e_0$ and, consequently [see (2.18)]:

$$\varkappa = \frac{2\sqrt{2}e_0}{\hbar c} H_{cm}\delta_0^2. \tag{2.19}$$

As can be seen from the calculations, the surface tension σ_{ns} is positive only for $\varkappa < 1/\sqrt{2}$. An analytical calculation of σ_{ns} encounters difficulties. In paper [29], this was only done for a sufficiently small $\varkappa$:

$$\sigma_{ns} = \frac{\delta_0 H_{cm}^2}{\sqrt{2} \cdot 3\pi\varkappa}, \quad \Delta = \frac{\sigma_{ns}}{H_{cm}^2/8\pi} = \frac{1.89\delta_0}{\varkappa}, \quad \sqrt{\varkappa} \ll 1. \tag{2.20}$$

From this, it is already seen that the Ψ-theory leads to σ_{ns} values of the required order of magnitude. It is only in paper [56] that the energy σ_{ns} is calculated analytically up to the terms of the order of $\varkappa\sqrt{\varkappa}$. The result is as follows [the value $\Gamma = 2\sqrt{2}/3$ corresponds to expression (2.20)]:

$$\sigma_{ns} = \frac{\delta_0 H_{cm}^2}{4\pi\varkappa} \Gamma, \quad \Gamma = \frac{2\sqrt{2}}{3} - 1.02817\sqrt{\varkappa} - 0.13307\varkappa\sqrt{\varkappa} + \dots \tag{2.21}$$

As $\varkappa$ increases, the energy σ_{ns} decreases and, in [29], it was pointed out that, according to numerical integration:

$$\sigma_{ns} = 0, \quad \varkappa = \frac{1}{\sqrt{2}}. \tag{2.22}$$

But it was also shown that for $\varkappa > 1/\sqrt{2}$, there occurs some specific instability of the normal phase, namely, nuclei of the superconducting phase are formed in it. Concretely, this instability arises in the field:

$$H_{c2} = \sqrt{2}\,\varkappa H_{cm}. \tag{2.23}$$

(It should be noted that (2.23) is present in [29] in an implicit form, and it was written explicitly in [57].) In the case $\varkappa < 1/\sqrt{2}$, the field H_{c2} corresponds to the limit of a possible supercooling of the normal phase (for $H < H_{c2}$, this phase becomes metastable; see also [57], where, as in some of my other papers, the field H_{c2} is denoted by H_{k1}). When $\varkappa > 1/\sqrt{2}$, it is clear from (2.23) that superconductivity is preserved in some form in the field $H > H_{cm}$ too and vanishes only in the field H_{c2}. Generally, it is just for $\varkappa = 1/\sqrt{2}$ that the change in the behavior of a superconductor becomes pronounced. Hence, there were no doubts in the validity of the result (2.22). Analytically this is proved, for example, in [30,37,38]. It turns out that for pure, superconducting metals we typically have $\varkappa < 1/\sqrt{2}$ or even $\varkappa \ll 1/\sqrt{2}$ (for instance, according to [30], $\varkappa$ is equal to 0.01 for Al, 0.13 for Sn, 0.16 for Hg, and 0.23 for Pb). Such superconductors are called type I superconductors. If $\varkappa > 1/\sqrt{2}$, the surface tension σ_{ns} is negative and we then deal with type II superconductors (for the most part alloys) whose behavior was first investigated thoroughly in experimental studies by L.V. Shubnikov[5] and co-authors as far back as 1935–36 (for references and explanations see [24, 58]). In [29], we considered only type I superconductors, and we read such a phrase there: 'For sufficiently large $\varkappa$, on the contrary, $\sigma_{ns} < 0$, which is indicative of the fact that such large $\varkappa$ do not correspond to the typically observed picture'. So we, in fact, overlooked the possibility of the existence of type II superconductors. Neither was I engaged in the study of type II superconductors later on. In this respect, I only made a remark in [57]. The theory of the behavior of type II superconductors based on the Ψ-theory was constructed in 1957 by A.A. Abrikosov [59] (see also [30,41]). As indicated in [59] and [30], p. 191, Landau was the first to suggest that in alloys $\varkappa > 1/\sqrt{2}$.

Allowing for (2.13) and (2.17), one can write:

$$H_{cm} = \left(\frac{4\pi (\alpha_c')^2}{\beta_c} \right)^{1/2} (T_c - T), \quad \delta_0 = \left(\frac{m_0 c^2 \beta_c}{16\pi e_0^2 \alpha_c'} \right)^{1/2} (T_c - T)^{-1/2}. \quad (2.24)$$

These expressions, the same as the whole Ψ-theory are, strictly speaking, valid only in the vicinity of T_c, i.e., the condition $(T_c - T) \ll T_c$ is needed. However, the condition of applicability of the theory for small $\varkappa$ is, in fact, more rigorous because to satisfy the local approximation, the penetration depth δ_0 must significantly exceed the size ξ_0 of the Cooper pair (the corresponding condition written in [30], Sect. 45 has the form $(T_c - T) \ll \varkappa^2 T_c$ but in [29] this, of course, could not yet be discussed). Along with the penetration depth δ_0, the Ψ-theory involves one more parameter which has the dimension of length – the so-called coherence length or the correlation radius (length):

$$\xi = \frac{\hbar}{\sqrt{2m_0|\alpha|}} = \frac{\hbar}{\sqrt{2m_0\alpha_c'(T_c - T)}} = \frac{\hbar \tau^{-1/2}}{\sqrt{2m_0\alpha_c'T_c}} = \xi(0)\tau^{-1/2}, \quad (2.25)$$

where $\tau = (T_c - T)/T_c$ and $\xi(0) = \hbar/\sqrt{2m_0\alpha_c'T_c}$ is a conditional correlation radius for $T = 0$ (we call it conditional because the Ψ-theory is, strictly speaking, applicable only in the vicinity of T_c). To compare the formulae written here with those of [30], one should bear in mind that in our expression (2.12) in [30] $m = 2m_0$ and, of course, $e = 2e_0$.

As is readily seen [see (2.18), (2.19), and (2.24)]:

$$\varkappa = \frac{m_0 c}{2e_0 \hbar} \sqrt{\frac{\beta_c}{2\pi}} = \frac{\delta_0(T)}{\xi(T)}. \quad (2.26)$$

---

[5] L.V. Shubnikov, a prominent experimental physicist, was guiltlessly executed in 1937.

In addition to these mentioned problems, some more points were considered in [29], namely the field in a superconducting half-space and critical fields for plates (films) in the case where superconductivity is destroyed by the field and current. The penetration depth of the field in a superconducting half-space adjoining a vacuum has the form:

$$\delta = \delta_0 \left[1 + f(\varkappa) \left(\frac{H_0}{H_{cm}} \right)^2 \right], \quad f(\varkappa) = \frac{\varkappa(\varkappa + 2\sqrt{2})}{8(\varkappa + \sqrt{2})^2} \tag{2.27}$$

where H_0 is the external field (the field for $z = 0$) and, by definition, $\delta = \int_0^\infty H(z)\,\mathrm{d}z/H_0$. The nonlinearity of the electrodynamics of superconductors, which was assumed already in [48] and is reflected in the dependence of δ on H_0, is fairly small. So, even for $\varkappa = 1/\sqrt{2}$ and $H_0 = H_{cm}$, the depth is $\delta = 1.07\delta_0$. In 1950, there were no accurate enough experimental measurements of $\delta(H)$. I am not sure that they have yet been carried out, though it is probable.

Now I should make or, rather, repeat one general remark. I was never long engaged in studying only superconductivity but researched various fields (see [1], p. 309 and [47] and Chap. 6 in [2]). As to the macroscopic theory of superconductivity (the Ψ-theory and its development), it was generally beyond the scope of my interest from a certain time (see Sect. 2.3). As a result, I am ignorant of the current state of the problem as a whole. Unfortunately, neither am I aware of the existence of a monograph compiling all the material (I am afraid there is no such book). Moreover, I forgot much of what I had done myself and now recollect the old facts, sometimes with surprise, when reading my own papers. That is why I cannot be convinced that my old calculations were unerring; I do not know the subsequent calculations and the results of their comparison with experiment. However, the present paper does not even claim to make a current review; it is only an attempt to elucidate some problems of the history of studies of superconductivity and superfluidity in an autobiographical context. Those uninterested will just not read it and, in this, I find some consolation.

The concluding part of paper [29] is devoted to a consideration of superconducting plates (films) of thickness $2d$ in an external magnetic field H_0 parallel to the film and also in the presence of a current $J = \int_{-d}^{+d} j(z)\,\mathrm{d}z$ (where $j(z)$ is the current density) flowing through the film. Instead of J, it is convenient to work in terms of the field $H_J = 2\pi J/c$ created by the current outside the film.

In the absence of current, the critical field H_c destroying superconductivity for thick films with $d \gg \delta_0$ is [see (2.27)]:

$$\frac{H_c}{H_{cm}} = 1 + \frac{\delta_0}{2d} \left(1 + \frac{f(\varkappa)}{2} \right), \quad d \gg \delta_0. \tag{2.28}$$

For sufficiently thin films, a transition to the normal state is a second-order transition (i.e., for $H_0 = H_c$, the function Ψ is equal to zero) and, for small

$\varkappa$, we have:

$$\left(\frac{H_{\mathrm{c}}}{H_{\mathrm{cm}}}\right)^2 = 6\left(\frac{\delta_0}{d}\right)^2 - \frac{7}{10}\varkappa^2 + \frac{11}{1400}\varkappa^4\left(\frac{d}{\delta_0}\right)^2 + \ldots, \quad d \ll \delta_0. \tag{2.29}$$

For films with a half-thickness $d > d_{\mathrm{c}}$, where:

$$d_{\mathrm{c}}^2 = \frac{5}{4}\left(1 - \frac{7}{24}\varkappa^2 + \ldots\right)\delta_0^2 \tag{2.30}$$

we are already dealing with first-order transitions with a release of latent transition heat (in other words, d_{c} is a tricritical point or, as it was termed before, a critical Curie point).

In the presence of a current and field (for $\varkappa = 0$):

$$\frac{H_{J_{\mathrm{c}}}}{H_{\mathrm{cm}}} = \frac{2\sqrt{2}}{3\sqrt{3}}\frac{d}{\delta_0}\left[1 - \left(\frac{H_0}{H_{\mathrm{c}}}\right)^2\right]^{3/2}, \quad d \ll \delta_0, \tag{2.31}$$

where H_{c} is the critical field for a given film in the absence of a current [see (2.29)], H_0 is the external field, and J_{c} is the critical current destroying superconductivity ($H_{J_{\mathrm{c}}} = 2\pi J_{\mathrm{c}}/c$).

The field H_{c} for such films is much larger than the critical field[6] H_{cm} for bulk samples and $H_{J_{\mathrm{c}}} \ll H_{\mathrm{cm}}$. It is interesting, however, that according to (2.29) and (2.31) (for $\varkappa = 0$ and $H_0 = 0$), we are led to

$$H_{\mathrm{c}}H_{J_{\mathrm{c}}} = \frac{4}{3}H_{\mathrm{cm}}^2. \tag{2.32}$$

In [29] we certainly tried to compare the theory with the then available experimental data. But, the latter was not numerous and, particularly importantly, their accuracy was low. To the best of my knowledge, all the results of the theory were later confirmed by experiment.

2.3 The Development of the Ψ-Theory of Superconductivity

In [29], neither did we solve all the problems, nor even those which were easy to formulate. Therefore, I naturally continued, although with some intervals, to develop the Ψ-theory for several years. For example, in paper [60] (see also [14]), I considered in more detail than in [29] the destruction of the superconductivity of thin films having half-thickness $d > d_{\mathrm{c}}$ [see (2.30)]: the condition $(\varkappa d/\delta_0)^2 \ll 1$ was used. Critical fields were found for supercooling

---

[6] The critical field for superconducting films was calculated with allowance for corrections of the order of $\varkappa^2$ in [280], where the theory was compared with the experiment.

and superheating. I note that not for films, but for cylinders and balls, critical fields were calculated (on the basis of the Ψ-theory) by V.P. Silin in [61] and myself in [62] (see also [229]). The critical current for superconducting films deposited onto a cylindrical surface was found in [63]. The question of normal phase supercooling [see (2.23)] was discussed in [57], which has already been mentioned, and the critical field for superheating of the superconducting phase in bulk superconductors was calculated in [62]. So, for a small $\varkappa$, the critical field for superheating (denoted as the field H_{k2} in [62]) is

$$\frac{H_{c1}}{H_{cm}} = \frac{0.89}{\sqrt{\varkappa}}, \quad \sqrt{\varkappa} \ll 1, \tag{2.33}$$

where the coefficient is obtained from numerical integration.[2*]

In several papers (see [14,32,55,64]), I discussed, in particular, the behavior of superconductors in a high-frequency field, but later on showed no interest in this issue and am now unaware whether these papers were of interest and importance for experiments (in respect of the behavior in a high-frequency field).

As I have already emphasized, the Ψ-theory can be immediately applied only in the vicinity of T_c. Naturally, I wished to extend the theory to the case of any temperature. In the framework of the phenomenological approach this goal can be achieved in different ways. So, Bardeen [65] suggested replacing the expression for the free energy F_{s0} from (2.12) with another expression involving a more complicated dependence of $F_{s0}\left(|\Psi|^2\right)$ on $|\Psi|^2$. The same object can, however, be attained [66] without the changing expression (2.12) but by assuming a certain dependence of the coefficients α and β on temperature, or, more precisely, on the ratio T/T_c. A somewhat different approach to the problem consists [67] not in assuming the dependence $F_{s0}\left(|\Psi|^2\right)$ in advance, but rather in finding it from comparison with the experiment.

After the creation of the BCS theory in 1957 and the papers [31] by Gorkov, I almost lost interest in the theory of superconductivity. Superconductivity was no longer an enigma (it had been an enigma for a long 46 years after its discovery in 1911). Quite a lot of other attractive problems existed, and I thought that I would drop superconductivity for ever. It was merely by inertia that, in 1959, when it became finally clear that the effective charge in the Ψ-theory was $e_{\text{eff}} \equiv e = 2e_0$, I compared [68] the Ψ-theory with the available experimental data and made sure that everything was all right. I will also mention the note [69] devoted to the allowance for pressure in the theory of second-order phase transitions as applied to a superconducting transition.

It was F. London [70] who had already pointed out that a magnetic flux through a hollow massive superconducting cylinder or a ring must be quantized, and that the flux quantum must be $\Phi_0 = hc/e$ and the flux $\Phi = k\Phi_0$, where k is an integer and e is the charge of the particles carrying the current. Naturally, London assumed $e = e_0$ to be a free electron charge. It was only in 1961 that the corresponding experiments were carried out (for references and a description of the experiments see, for example, [71]) demonstrating that, in fact, $e = 2e_0$. The latter is quite clear from the point of view of the BCS theory according to which it is pairs of electrons that are carried over. Thus

$$\Phi = \frac{hck}{2e_0} = \frac{\pi \hbar c k}{e_0} = \Phi_0 k, \quad \Phi_0 = 2 \times 10^{-7} \, \text{G cm}^2 \quad (k = 0, 1, 2, \ldots). \quad (2.34)$$

This result (2.34) refers, however, only to the case of doubly connected bulk samples, for instance, hollow cylinders with wall thicknesses substantially exceeding the magnetic field penetration depth δ in a superconductor. And yet, samples of any size, as well as those located in an external magnetic field, etc., are also of interest. Within the framework of the Ψ-theory, I solved this problem in paper [72]. A similar but less thorough and comprehensive analysis appeared nearly simultaneously in [73, 74] (all the papers [72–74] were submitted for publication in mid-1961).

I have not yet mentioned my papers [75] and [76], which were written before the creation of the BCS theory which, however, fell out of the scope of direct application of the Ψ-theory [29]. So, in [75], the Ψ-theory was extended to the case of anisotropic superconductors. In the 'low-temperature' (conventional) superconductors known at that time, anisotropy is either absent altogether (isotropic and cubic materials) or is fairly small. It was apparently for this reason that in [29] we assumed, even without reservations, that metals are isotropic. But as early as in paper [22], when I considered thermoelectric phenomena, I had to examine an anisotropic (i.e., non-cubic) crystal and, in view of this, I generalized the London theory (2.4), (2.5) by introducing a symmetric tensor of rank two, Λ_{ik}, instead of the scalar Λ so that rot $\mathbf{\Lambda}(\mathbf{j}) = -\mathbf{H}/c$, $\Lambda_i(\mathbf{j}) = \Lambda_{ik} j_k$, (here $\mathbf{j} = \mathbf{j}_s$ is the superconducting current density). Such a generalization is, of course, obvious enough but I mention it here because in the extensive review [20] Bardeen refers in this connection only to papers [78, 79] by M. Laue which appeared later.

In [75], the complex scalar function $\Psi(\mathbf{r})$ for anisotropic material is introduced as before but the free energy is written not in the form (2.12) but as

$$F_{sH} = F_{s0} + \frac{H^2}{8\pi} + \frac{1}{2m_k} \left| -\mathrm{i}\hbar \frac{\partial \Psi}{\partial x_k} - \frac{2e_0}{c} A_k \Psi \right|^2, \quad (2.35)$$

where doubly occurring indices are summed and, in [75], the charge e is taken instead of $2e_0$, and, for an isotropic or cubic material, $m_1 = m_2 = m_3 = m$, and we obtain (2.12).

As mentioned previously, the anisotropy in 'conventional' superconductors is not large, i.e., the 'effective masses' m_k little differ from one another. But, in the majority of high-temperature superconductors, in contrast, the anisotropy is very large and it is (2.35) and the corollaries to it, partially mentioned already in [75], that are widely used. An interesting effect related to the anisotropy of a superconductor is noted in [238].

Among the superconductors known in the 1950s, there was not a single ferromagnetic. This is, of course, not accidental. The point is that even digressing from microscopic reasons, the presence of ferromagnetism hampers the occurrence of superconductivity [76]. Indeed, one can see that in the depth of a ferromagnetic superconductor the magnetic induction $\boldsymbol{B}$ must also be zero. However, spontaneous magnetization $\boldsymbol{M}_s$ causes induction $\boldsymbol{B} = 4\pi\boldsymbol{M}_s$. Consequently, in a ferromagnetic superconductor, even in the absence of an external magnetic field, there must flow a surface superconducting current compensating for the 'molecular' current responsible for magnetization. From this, it follows that a thermodynamic critical magnetic field for a ferromagnetic superconductor is

$$H_{\mathrm{cm}}(T) = \frac{H_{\mathrm{cm}}^{(0)}(T)}{\sqrt{\mu}} - \frac{4\pi M_s}{\mu}, \quad H_{\mathrm{cm}}^{(0)} = \sqrt{8\pi(F_{\mathrm{n0}} - F_{\mathrm{s0}})}, \quad (2.36)$$

where the ferromagnetic is assumed to be 'ideal', i.e., for it $\boldsymbol{B} = \boldsymbol{H} + 4\pi\boldsymbol{M} = \mu\boldsymbol{H} + 4\pi\boldsymbol{M}_s$ (μ is magnetic permittivity) and F_{n0} and F_{s0} are free energies for the normal and superconducting phases of a given metal in the absence of magnetization and a magnetic field. Obviously, superconductivity is only possible under the condition $H_{\mathrm{cm}}^{(0)}(0) > 4\pi M_s/\sqrt{\mu}$ which can hold, in fact, only for ferromagnetics with a not very large spontaneous magnetization M_s. With the appearance of the BCS theory, it became clear that superconductivity and ferromagnetism obstruct each other, even irrespective of the previously mentioned so-called electromagnetic factor. Indeed, conventional superconductivity is associated with the pairing of electrons with oppositely directed spins, while ferromagnetism corresponds to parallel spin orientation. Thus, the exchange forces that lead to ferromagnetism obstruct the appearance of superconductivity. Nevertheless, ferromagnetic superconductors were discovered, but naturally with fairly low values of T_c and the Curie temperature T_M (see [77, 217, 266] and also Chap. 6 in [2]).

Unfortunately, I am not aware of corresponding experiments and wish to emphasize here that the 'electromagnetic factor' was allowed for in only the simplest, trivial case of an equilibrium uniform magnetization of bulk metal. However, there exist alternative possibilities [76] (see also [230]).

For example, let us assume that a ferromagnetic metal possesses a large coercive force and that in the external field $H_c < H_{\mathrm{coer}}$ magnetization can remain directed opposite to the field (for simplicity, we consider cylindrical samples in a parallel field). Then, for $M_s < 0$ (the magnetization is directed oppositely to the field), superconductivity may exist under the condition $H_{\mathrm{cm}}^{(0)}(0) > 4\pi|M_s|/\sqrt{\mu} - \sqrt{\mu}H_{\mathrm{coer}}$, i.e., in principle, the 'electromagnetic factor' may be absolutely insignificant. Of even greater interest are possibilities arising in the case of thin films and generally small-size samples. For them, the critical field $H_c^{(0)}$, as is well known and has already been mentioned, may substantially exceed the field $H_{\mathrm{cm}}^{(0)}$ for bulk metal. At the same time, a critical field for a ferromagnetic superconducting film, even for $M_s > 0$ (when the magnetization is directed along the field), has, as before, the form (2.36) but with $H_{\mathrm{cm}}^{(0)}$ replaced by $H_c^{(0)}$. Now, the presence of magnetization M_s may already be of no importance. Thus, additional possibilities open up for investigating ferromagnetic superconductors. I do not know if these possibilities have ever been considered.[7]

We have up to now discussed only equilibrium or metastable (superheated or supercooled) states of superconductors, fluctuations being totally ignored. Meanwhile, fluctuations near phase transition points, especially for second-order transitions, generally speaking, play an important role (see, for example, [34], Sect. 146). In the case of superconductors, one should expect fluctuations of the order parameter Ψ both below and above T_c. I can tell the reader about my activity in this field. In 1952, at the end of [80], it was noted that fluctuations of the 'concentration of superconducting electrons' n_s must also be present above T_c and that this must affect, first of all, the complex dielectric constant of a metal. At the end of review [14], this remark was made again, with an emphasis on the fact that as $T \to T_c$ the fluctuations must be large. However, I never elaborated upon this observation later. Fourteen years had passed before V.V. Schmidt [81] (whose untimely death occurred in 1985) went farther and considered (with a reference to [80]) the question of the fluctuational specific heat of small balls above T_c, and also mentioned the possibility of observing the fluctuational diamagnetic moment of such balls. It is curious that another two physicists with this name investigated [82, 83] the same issue and, moreover, considered fluctuational conductivity above T_c (for the fluctuation effects see also [30, 84, 85]).

Let us now turn to a very important question of the applicability limits of Landau's phase transition theory, both in the general context and in its application to superconductors [86].

Landau's phase transition theory [34, 50] is well known to be the mean field theory (or, as it is sometimes referred to, the molecular or self-consistent field theory). This means that the free energy (or a corresponding thermodynamic potential) of the type:

$$F = F_0 + \alpha \eta^2 + \frac{\beta}{2}\eta^4 + \frac{\gamma}{6}\eta^6 + g(\nabla \eta)^2 \qquad (2.37)$$

does not include the contribution from the fluctuations of η.

As we have seen in the example of a superconductor, when $\eta = \Psi$ [see (2.12), (2.13)], below the second-order transition point (we set $\gamma = 0$), the equilibrium value is

$$\eta_0^2 = -\frac{\alpha}{\beta} = \frac{\alpha'_c(T_c - T)}{\beta_c}. \qquad (2.38)$$

Taking the Landau theory as the first approximation and using it as a basis, one can find the fluctuations of various quantities, in particular, the parameter η itself. Naturally, the Landau theory holds true and the fluctuations calculated on its basis hold true only as long as they are small compared to

---

[7] The superconductivity in ferromagnetics has attracted much attention [?, 253–255, 258, 273, 274, 276].

the mean values obtained within the Landau theory. In application to η, this means that the condition

$$\overline{(\Delta\eta)^2} \ll \eta_0^2 \tag{2.39}$$

must hold, where obviously $\overline{(\Delta\eta)^2}$ is the statistical mean of the fluctuation of the quantity η (the fluctuation $\overline{(\Delta\eta)}$ is zero because we calculate the deviations from the value η_0 corresponding to the minimum free energy).

The use of criterion (2.39) leads to the following condition of applicability of the Landau theory (see [86–88]):

$$\tau \equiv \frac{T_c - T}{T_c} \gg \frac{k_B^2 T_c \beta_c^2}{32\pi^2 \alpha_c' g^3} \tag{2.40}$$

where k_B is the Boltzmann constant. This means that the Landau theory can be exploited within the temperature range in the vicinity of the transition point T_c satisfying inequality (2.40). A condition of type (2.40) or similar was derived in different but close ways in [34, 86–88]. For example, in [88] the condition of applicability of the Landau theory is written in the form (in our notation; moreover, in [34, 88] k_B was set unity):

$$\mathrm{Gi} = \frac{T_c \beta_c^2}{\alpha_c' g^3} \ll \tau \ll 1, \quad \tau = \frac{T_c - T}{T_c}. \tag{2.41}$$

The number Gi in [88] was called the Ginzburg number but I never employ this terminology for the reason mentioned earlier in respect to the Ψ-theory. In my opinion it is more appropriate to employ a criterion of the form (2.40) because the coefficient $1/32\pi^2$ is fairly small and this extends, in fact, the limits of applicability of the Landau theory (note that in [86] the coefficient $1/32\pi^2$ in the final expression (2.5b) is omitted but it is clear from (2.4) for $\overline{(\Delta\eta)^2}$).

Obviously, the smaller the number Gi is, the closer to the transition point the Landau theory can be used, in which, in particular, the specific heat simply undergoes a jump (without λ-singularity) and $\eta_0^2 \sim (T_c - T)$. This immediately implies, for example, that in liquid helium (^{4}He) the parameter Gi is large and this results in the existence of the λ-singularity. In [86], various transitions are discussed, the most detailed consideration being given to ferroelectrics to which the Landau theory is generally well applicable, as to other structure phase transitions. This subject was discussed many years later in paper [89] but we shall not touch upon it here; see Chap. 5 in [2]. In the present paper, we are concerned with superconducting transitions and the λ-transition in liquid helium. The latter is dealt with in Sect. 2.4. As far as superconductors are concerned, from comparison of the expressions in (2.12) with $e = 2e_0$ and $m = m_0$, (2.25), (2.26), (2.37), and (2.40), it follows that condition (2.40) takes on the form

$$\tau \equiv \frac{T_c - T}{T_c} \gg \tau_G \equiv \frac{(k_B \beta_c)^2}{32\pi^2 (\alpha_c')^4 T_c^2 [\xi(0)]^6}. \tag{2.42}$$

This expression, however, bears no specific features for superconductors and refers to any second-order transition described by the Landau theory. In the framework of this theory, as is clear from [34] and, for example, from (2.13) or (2.37), the jump ΔC of specific heat $C = T\,dS/dT$, where $S = -\partial F/\partial T$ is entropy, at transition is

$$\Delta C = \frac{(\alpha_c')^2 T_c}{\beta_c}. \tag{2.43}$$

From (2.43), it is clear that condition (2.42) involves, in particular, the directly measurable quantity ΔC. Next, for superconductors [see (2.13), (2.23), (2.25), (2.26), and (2.34)],

$$H_{cm}^2 = \frac{4\pi(\alpha_c')^2}{\beta_c}(T_c - T)^2 = \frac{4\pi(\alpha_c')^2 T_c^2}{\beta_c}\tau^2 \equiv H_{cm}^2(0)\tau^2,$$

$$H_{c2}^2 = 2\varkappa^2 H_{cm}^2, \quad \xi^2 = \frac{\hbar^2}{2m_0\alpha_c' T_c}\tau^{-1} \equiv \xi^2(0)\tau^{-1}, \tag{2.44}$$

$$\varkappa^2 = \frac{m_0^2 c^2 \beta_c}{8\pi e_0^2 \hbar^2}, \quad \xi^{-2}(0) = \frac{2e_0}{\hbar c}H_{c2}(0) = \frac{2\pi H_{c2}(0)}{\Phi_0}, \quad H_{c2}^2(0) = 2\varkappa^2 H_{cm}^2(0).$$

To avoid misunderstanding, we shall stress that all our consideration, as well as the Ψ-theory itself, refers directly only to the region near T_c. Consequently, the quantities $H_{cm}(0)$ and $H_{c2}(0)$ are somewhat formal and are not at all the true values of the fields $H_{cm}(T)$ and $H_{c2}(T)$ at $T = 0$. In view of this, it would be more correct to employ the derivatives $(dH_{cm}/dT)_{T=T_c} = -H_{cm}(0)/T_c$ and $(dH_{c2}/dT)_{T=T_c} = -H_{c2}(0)/T_c$ which can be measured in the experiment.

Allowing for (2.43) and (2.45), one can rewrite condition (2.42) in the form

$$\tau \gg \tau_G = \left(\frac{2\pi}{\Phi_0}\right)^3 \frac{H_{c2}^3(0)}{32\pi^2(\Delta C)^2}, \quad \Phi_0 = \frac{\pi\hbar c}{e_0}. \tag{2.45}$$

For type I superconductors, the substitution in (2.42) and (2.45) of the values of $\xi(0)$ (or $H_{c2}(0)$) and ΔC known from the experiment, even without account of the factor $1/32\pi^2 \sim 3 \times 10^{-3}$, yields the estimate $\tau_G \sim 10^{-15}$ (see [86] for $T_c \sim 1$ K) or, on the basis of the BCS model, the estimate $\tau_G \sim (k_B T_c/E_F)^4 \sim 10^{-12}$–$10^{-16}$ (here E_F is the Fermi energy; see [30], Sect. 45; 88). Physically, it is obvious that the smallness of the value τ_G for superconductors is due to the high value of the correlation radius $\xi(0)$ in type I superconductors. In this case, the characteristic value $\xi(0) \sim \xi_0 \sim 10^{-4}$–$10^{-5}$cm is of the order of the size of a Cooper pair. For structure phase transitions, $\xi(0) \sim d \sim 3 \times 10^{-8}$cm and is of the order of interatomic length and the fluctuation region must be seemingly large. But, in this case (in particular, in ferroelectrics), the relative smallness of τ_G is caused by other factors (see [86,89]).

Thus, the Ψ-theory is, generally speaking, well applied to superconductors. The words 'generally speaking' refer to several circumstances. Firstly, we have considered here the three-dimensional case. For quasi-two-dimensional (thin films), quasi-one-dimensional (thin wires, etc.), and quasi-zero-dimensional (small seeds, say, balls) superconductors, the conditions of applicability of the theory are different: the fluctuation region is wider than for a three-dimensional system. Unfortunately, I do not know all aspects of the problem (see, however, [90]). Secondly, as has already been emphasized, good applicability of the mean field approximation (the Landau theory and, in particular, the Ψ-theory) is in no way an obstruction to the calculation of various fluctuation effects, as long as they are sufficiently small (see, for example, [81–85, 90, 91]). It is of importance, especially in application to high-temperature superconductors (HTSCs), that paper [90] analyses, on the basis of (2.35), the anisotropic case. Third, in a number of superconductors (dirty alloys, HTSCs), the parameter $\varkappa$ is large or even very large (reaching hundreds) while the correlation length is small. Then the fluctuation region, i.e., the temperature range in which inequality (2.42), (2.45) is violated, is not so small. So, in [90], we present the values $\tau_{\mathrm{G}} = (0.2\text{–}2) \times 10^{-4}$ for HTSCs. Somewhat lower values are reported in [92]. For $\tau_{\mathrm{G}} \sim 10^{-4}$ and $T_{\mathrm{c}} \sim 100\,\mathrm{K}$, the width of the fluctuation region is $\Delta T \sim 10^{-2}\,\mathrm{K}$ (in this region the fluctuations are already high and are, therefore, not a small correction). This region does not seem to be so very large, but in experiments the variation of the specific heat of some HTSCs near T_{c} has a clearly pronounced λ-shaped form similar to the one we observe in Helium II (see [93], p. 2; 132), where the original literature is cited).

In view of the latter circumstance, it seems interesting to extend the Ψ-theory to the fluctuation region. We shall touch upon this issue in Sect. 2.4 because this extension was proposed in the application to liquid helium. But after the discovery of HTSCs in 1986–1987, such a 'generalized Ψ-theory' was suggested in the application to superconductors as well [54, 90, 94].

Underlying the 'generalized' Ψ-theory of superconductivity is, for instance, the following expression:

$$
\widetilde{F} = \widetilde{F}_{\mathrm{n}0} + \frac{C_0 T_{\mathrm{c}}}{2} \tau^2 \ln \tau + \int \left[-a_0 \tau^{4/3} |\Psi|^2 + \frac{b_0}{2} \tau^{2/3} |\Psi|^4 \right.
$$
$$
\left. + \frac{g_0}{3} |\Psi|^6 + \frac{\hbar^2}{4 m_k} \left| \left(\nabla_k - \mathrm{i} \frac{2 e_0}{\hbar c} A_k \right) \Psi \right|^2 \right] \mathrm{d}V \qquad (2.46)
$$

for the free energy which leads to the following equation for Ψ:

$$
- \frac{\hbar^2}{4 m_k} \left(\nabla_k - \mathrm{i} \frac{2 e_0}{\hbar c} A_k \right)^2 \Psi + \left(-a_0 \tau^{4/3} + b_0 \tau^{2/3} |\Psi|^2 + g_0 |\Psi|^4 \right) \Psi = 0. \quad (2.47)
$$

If one neglects anisotropy and sets $m_k = m_0/2$, then (2.47) will differ from (2.14) by a transformed temperature dependence of the coefficients and by the presence of the term proportional to $|\Psi|^4 \Psi$. Taking the example of helium II, we shall see in Sect. 2.4 that the 'generalized' Ψ-theory entails a number of consequences near T_{c} which correspond in reality in the case of liquid helium.

One might think that this could also be extended to superconductors with a very small correlation length. Such a case corresponds in a certain measure to the Schafroth model [16] which involves small-sized pairs. One of the directions of HTSC theory is based precisely on this model [93].

Another example of generalization of Ψ-theory near the transition point can be seen in [277].

An important point in the 'generalized' Ψ-theory is the problem of boundary conditions. Condition (2.15) is, generally speaking, already insufficient here and should be replaced [37, 90, 95] by a more general condition:

$$n_k \Lambda_k \left[\frac{\partial \Psi}{\partial x_k} - \mathrm{i} \frac{2e_0}{\hbar c} A_k \Psi \right] = -\Psi \tag{2.48}$$

on the boundary with a vacuum or a dielectric, where all the quantities are, of course, taken on the boundary, n_k are the components of the unit vector $\boldsymbol{n}$ perpendicular to the boundary, and Λ_k are some coefficients having dimensions of length, sometimes referred to as extrapolation lengths. For the isotropic case, when $\Lambda_k = \Lambda$, (2.48) takes on the form:

$$\boldsymbol{n} \left(\nabla \Psi - \mathrm{i} \frac{2e_0}{\hbar c} \boldsymbol{A} \Psi \right) = -\frac{1}{\Lambda} \Psi \tag{2.49}$$

[this Λ should not be confused with the coefficient (2.17) involved in the London theory (2.4), (2.5)].

For $\Lambda_k \gg \xi_k(T)$, condition (2.49) becomes (2.15) because, generally speaking, $\partial \Psi / \partial x_k \sim \Psi / \xi_k$. In the case $\Lambda_k \ll \xi_k(T)$, however, we arrive at the boundary condition

$$\Psi = 0. \tag{2.50}$$

This condition on a rigid wall was chosen in the initial Ψ-theory of superfluidity [94, 96]. As far as I know, the 'generalized' Ψ-theory of superconductivity was never used after paper [90]. Two reasons for this are possible. On the one hand, the 'generalized' Ψ-theory has no reliable microscopic grounds (as distinct from the conventional Ψ-theory of superconductivity considered earlier). On the other hand, the investigations of HTSC are obviously at such a stage now that it has probably not yet become necessary to solve problems requiring the application of the 'generalized' Ψ-theory. As far as the conventional Ψ-theory is concerned, its application to HTSC is also now only rather small scale.

It should be remarked that the Ψ-theory of superconductivity [29] might be, and sometimes has to be, generalized in view of introducing a more complex Ψ-functions. In paper [275], for instance, the author considered a generalization of the Ψ-theory in the context of MgB_2 superconductivity by introducing two functions Ψ_1 and Ψ_2 (so that the order parameter has now the Ψ_1 and Ψ_2 components [275]; see also [277]).

I have dwelt on the development of the initial Ψ-theory [29] in three directions: allowing for anisotropy [75], for ferromagnetic superconductors [76], and in a fluctuation region [90]. Also of importance are extensions in another two directions, namely to the non-stationary case, when the function Ψ is time dependent, and to superconductors in which the order parameter is not reduced to the scalar complex function $\Psi(\boldsymbol{r})$. I obtained no results in either of these two directions. True, in what concerns the non-stationary generalization of the Ψ-theory, I already understood [64] in 1950 that this task did exist, but restricted myself to the remark that (2.14) might be supplemented with the term $i\hbar\partial\Psi/\partial t$. Meanwhile, an allowance for relaxation is more significant. The corresponding equations for $\Psi(\boldsymbol{r},t)$ are discussed in reviews [85, 97, 237]. As to the so-called 'unconventional' superconductors in which Cooper (or analogous) pairs are not in the s-state, I not only failed to contribute to this field, I also have a poor knowledge of it. By the way, the possibility of 'unconventional' pairing was first pointed out [98] for superfluid ^{3}He, and this fact was later confirmed. In the case of superconductivity, the 'unconventional' pairing takes place for at least several superconductors with heavy fermions (UB$_{13}$, CeCu$_2$Si$_2$, UPt$_3$) and, apparently, several HTSCs – the cuprates. I shall restrict myself only to pointing to one of the pioneering papers in this field [99] and reviews [100–103]. It is a pleasure to me to note also that 'unconventional' superconductors are now the subject of successful research by Y.S. Barash [104], my immediate colleague (our joint research was, however, conducted in quite a different field – the theory of Van der Waals forces [105]). It is noteworthy that an appropriately extended Ψ-theory is extensively used for 'unconventional' superconductors as well [99–102].[9*,10*]

2.4 The Ψ-Theory of Superfluidity

As I have already mentioned, the behavior of liquid helium near the λ-point was beyond the scope of Landau's interests. He also remained indifferent to the behavior of superfluid helium near a rigid wall. As for me, I was for some reason interested in both these questions from the very beginning of my work in the field of superfluidity, i.e., from 1943 on [19]. I have already mentioned the attempt [49] to introduce the order parameter ρ_s near the λ-point. As regards the behavior of helium near the wall, it looks like this. Helium atoms stick to the wall (they wet it, so to say). How can it be combined with a flow along the wall of the superfluid part of the liquid with density ρ_s and a velocity $\boldsymbol{v}_s$? We know that in the Landau theory of superfluidity [4] the velocity $\boldsymbol{v}_s$ along the wall (as distinct from the velocity $\boldsymbol{v}_n$ of a normal liquid) does not become zero on the wall. This means that, on the wall, the velocity $\boldsymbol{v}_s$ must become discontinuous (the velocity $\boldsymbol{v}_s$ cannot tend gradually to zero because of the condition rot $\boldsymbol{v}_s = 0$). This velocity discontinuity must be associated with a certain surface energy σ_s [106]. Estimates show that this energy σ_s is rather high ($\sigma_s \sim 3 \times 10^{-2}\mathrm{erg\,cm^{-2}}$) and its existence must have led to a pronounced effect. Specifically, something like dry friction must have been observed – to move a rigid body placed in Helium II, the energy

$\sigma_s S$ must have been expended, where S is the body (say, plate) surface area. However, specially conducted experiments showed [107] that no energy $\sigma_s S$ is actually needed and a possible value of σ_s is at least, by many orders of magnitude, smaller, than the previously mentioned estimates [106]. How can this contradiction be eliminated? The solution of the problem I saw in the assumption that the density ρ_s decreases on approaching the wall and, on the wall itself, $\rho_s(0) = 0$. Thus, the discontinuity of the velocity $\boldsymbol{v}_s$ on the wall is of no importance because the flow $\boldsymbol{j}_s = \rho_s \boldsymbol{v}_s$ tends gradually to zero on the wall itself even without a change of velocity $\boldsymbol{v}_s$. By that time (1957), the Ψ-theory of superconductivity [29] had long since been constructed and there was no problem in extending it to the case of superfluidity and with the boundary condition $\Psi(0) = 0$ on the wall [see (2.50)], which provided the condition $\rho_s(0) = 0$, as well.

Unfortunately, I do not at all remember how far I had advanced in constructing the Ψ-theory of superfluidity before I learnt that L.P. Pitaevskii was engaged in solving the same problem. We naturally joined our efforts and the outcome was our paper [96] which we submitted for publication on 10 December, 1957.

The Ψ-theory of superfluidity constructed in [96] will, henceforth, be referred to as the initial Ψ-theory of superfluidity. The point is that this theory was later found to be inapplicable to Helium II in the quantitative respect and we had to generalize it. Such a generalized Ψ-theory of superfluidity, developed by A.A. Sobyanin and myself [108–112], is far from being so well grounded as the Ψ-theory of superconductivity. In this connection, and, I think, in view of an insufficient awareness of the distinction between the generalized theory and the initial one [96], the Ψ-theory of superfluidity has not attracted attention and, at the present time, remains undeveloped[8] and not systematically verified. Meanwhile, the microtheory of superfluidity is not nearly so well developed as the microtheory of superconductivity, and the role of the macrotheory of superfluidity is particularly high. This has led Sobyanin and myself to the conviction that the development of the Ψ-theory of superfluidity and its comparison with an experiment would be highly appropriate.

The most comprehensive of the cited reviews devoted to the generalized Ψ-theory of superfluidity [110] amounts to 78 pages. This alone makes it clear that, in this article, I have no way of giving an in-depth consideration to the Ψ-theory of superfluidity. Here I shall restrict myself to brief remarks.

We shall begin with the initial theory [96]. It is constructed in much the same manner as the Ψ-theory of superconductivity [29]. As the order parameter, we took the function $\Psi = |\Psi| \exp i\varphi$ acting as an 'effective wavefunction of the superfluid part of a liquid' and so the density ρ_s and the velocity $\boldsymbol{v}_s$ are expressed as

---

[8] One of the reasons, and perhaps even the main one, is the fact that A.A. Sobyanin has become a politician and for several years now has not been working practically as a physicist.[3*]

$$\rho_s = m|\Psi|^2, \quad v_s = \frac{\hbar}{m}\nabla\varphi,$$

$$j_s = \rho_s v_s = -\frac{i\hbar}{2}(\Psi^*\nabla\Psi - \Psi\nabla\Psi^*) = \hbar|\Psi|^2\nabla\varphi, \qquad (2.51)$$

where $m = m_{He}$ is the mass of a helium atom and a convenient normalization of Ψ is chosen; in [96] it is shown (see also later) that in the expression for v_s we have $m = m_{He}$, irrespective of the manner in which Ψ is normalized. Then there come the expressions

$$F = F_0 + \frac{\hbar^2}{2m}|\nabla\Psi|^2,$$

$$F_0 = F_I + \alpha|\Psi|^2 + \frac{\beta}{2}|\Psi|^4, \quad \alpha = \alpha'_\lambda(T - T_\lambda), \quad \beta = \beta_\lambda \qquad (2.52)$$

which are usual for the mean field theory (the Landau phase transitions theory), where $F_I(\rho, T)$ is the free energy of Helium I and T_λ is the temperature of the λ-point. In equilibrium, homogeneous Helium II

$$|\Psi_0|^2 = \frac{\rho_s}{m} = \frac{|\alpha|}{\beta_\lambda} = \frac{\alpha'_\lambda(T_\lambda - T)}{\beta_\lambda}, \quad \Delta C_p = C_{p,II} - C_{p,I} = T_\lambda\frac{(\alpha'_\lambda)^2}{\beta_\lambda}. \qquad (2.53)$$

In inhomogeneous Helium II, the function Ψ obeys the equation:

$$-\frac{\hbar^2}{2m}\nabla\Psi + \alpha\Psi + \beta_\lambda|\Psi|^2\Psi = 0 \qquad (2.54)$$

which should be solved with the boundary condition (2.50) on a rigid wall.

As in (2.25), we introduce the correlation length (in [96] it is denoted by l):

$$\xi(T) = \frac{\hbar}{\sqrt{2m|\alpha|}} = \frac{\hbar\tau^{-1/2}}{\sqrt{2m\alpha'_\lambda T_\lambda}} = \xi(0)\tau^{-1/2}, \quad \tau = \frac{T_\lambda - T}{T_\lambda} = \frac{t}{T_\lambda}. \qquad (2.55)$$

The estimate presented in [96] and based on the data of ΔC_p and ρ_s measurements [see (2.54)] gives approximately $\xi(0) \sim 3 \times 10^{-8}$cm. At the same time, the Ψ-theory is applicable, provided only that the macroscopic Ψ-function changes little on atomic scales. This implies the condition $\xi(T) \gg a \sim 3 \times 10^{-8}$cm (here a is the mean interatomic distance in liquid helium). Consequently, the Ψ-theory can only hold near the λ-point for $\tau \ll 1$, say, for $(T_\lambda - T) < (0.1$–$0.2)$ K. Of course, proximity to T_λ is also the condition of applicability of expansion (2.52) in $|\Psi|^2$. The small magnitude of the length $\xi(0)$ in helium leads at the same time to considerable dimensions of the fluctuation region [86]. Indeed, applying criterion (2.42), we arrive at the value $\tau_G \sim 10^{-3}$ for helium (see [108], formula (2.46)). Thus, it turns out that the initial Ψ-theory of superfluidity can only hold under the condition 10^{-3}K $\ll (T_\lambda - T) \lesssim 0.1$ K, i.e., it is practically inapplicable because in the studies of liquid helium, of particular interest is exactly the range of values

$(T_\lambda - T) \ll 10^{-3}$K. The fact that the mean field theory leading to the jump in specific heat (2.53) does not hold for liquid helium (we certainly mean ^{4}He) is attested by the existence of a λ-singularity in the specific heat, as well as the circumstance that the density ρ_s near T_λ does not behave at all proportional to $(T_\lambda - T)$, i.e., according to (2.53) but rather changes by the law

$$\rho_s(\tau) = \rho_{s0}\tau^\zeta, \quad \zeta = 0.6705 \pm 0.0006 \tag{2.56}$$

where the value of ζ is borrowed from the most recently reported data [113]. Note that, in [108], we gave the value $\zeta = 0.67 \pm 0.01$ and, in [110], the values $\zeta = 0.672 \pm 0.001$ and $\rho_s = 0.35\tau^\zeta\,\mathrm{g\,cm}^{-3}$. Hence, to a high accuracy, we have

$$\zeta = \frac{2}{3}. \tag{2.57}$$

I cannot judge whether ζ actually differs from 2/3 but if it does, the difference does not exceed 1%. It is noteworthy that, in 1957, when paper [96] was accomplished, the variation of ρ_s by the law (2.56) was not yet known. We, therefore, did not raise an alarm immediately (the λ-type behavior of specific heat is less crucial in this respect because it may not be associated with variations of Ψ, whereas the density ρ_s is proportional to $|\Psi|^2$).

Thus, the initial Ψ-theory of superfluidity [96] is inapplicable to liquid helium (^{4}He). However, owing to its simplicity, it has a qualitative and, occasionally, even quantitative significance for ^{4}He as well. The main thing is that liquid ^{4}He is not the only existing superfluid liquid, suffice it to mention liquid ^{3}He at very low temperatures, ^{3}He–^{4}He solutions, non-dense ^{4}He films, and neutron liquid in neutron stars, as well as possible superfluidity in an exciton liquid in crystals, in supercooled liquid hydrogen [114], and in the Bose–Einstein condensate of the gas of various atoms (it is this very question that is presently commanding the attention of physicists; see, for example, [115] and references therein).[4*] In some of these cases, the fluctuation region may appear to be small enough, so that the initial Ψ-theory of superfluidity may prove sufficient. This is apparently the situation in the particularly important case of superfluidity in ^{3}He (see note 8). We shall, therefore, dwell briefly on the results obtained in [96].

We found the distribution $\rho_s(z)$ near a rigid wall and in a liquid helium film of thickness d. The function $\Psi(z)$ and, of course, $\rho_s = m|\Psi|^2$, where z is the coordinate perpendicular to the film, has a dome-like shape because on the boundaries of the film we have $\Psi(0) = \Psi(d) = 0$ [see (2.50)]. Naturally, for a sufficiently small thickness d, the equilibrium value is $\Psi = 0$, i.e., the superfluidity vanishes. The corresponding critical value d_c (for $d < d_c$ a film is not superfluid) is equal to

$$d_c = \pi\xi(T) = \frac{\pi\hbar\tau^{-1/2}}{\sqrt{2m\alpha'_\lambda T_\lambda}}, \quad \tau = \frac{T_\lambda - T}{T_\lambda}. \tag{2.58}$$

This result implies that, for a film, the λ-transition temperature is lower than that for 'bulk' helium. Concretely, from (2.58), it follows that, for a film, the

λ-transition takes place at a temperature $(T_\lambda \equiv T_\lambda(\infty))$:

$$T_\lambda(d) = T_\lambda - \frac{\pi^2 \hbar^2}{2m\alpha'_\lambda d^2} = T_\lambda - \frac{\pi^2 T_\lambda \xi^2(0)}{d^2}. \qquad (2.59)$$

The specific heat of the film changes with varying d, too. Such effects in small samples are observed experimentally. In [96] we also solved the problem of the vortex line, the value of Ψ on its axis being equal to zero and the velocity circulation around the line being

$$\oint \boldsymbol{v}_\mathrm{s}\, \mathrm{d}\boldsymbol{s} = \frac{2\pi\hbar k}{m}, \quad k = 0, 1, 2, \ldots \qquad (2.60)$$

In this formula, the ^{4}He atom mass $m = m_\mathrm{He}$ should be used considering that the circulation cannot change with temperature and, as was shown by Feynman [116], at $T = 0$ it is the mass m_He that enters into (2.60). Finally, in [96], we found the surface energy on the boundary between Helium II and a rigid body and the vortex line energy.

The fact that for liquid helium and a number of other transitions, the mean field (Landau theory) does not hold led to the appearance of the generalized theory in which the free energy is written in the form (2.37) but with a different temperature dependence of the coefficients. Specifically, for the order parameter Ψ, we write

$$F_\mathrm{II} = F_\mathrm{I} - a_0\tau|\tau|^{1/3}|\Psi|^2 + \frac{b_0}{2}|\tau|^{2/3}|\Psi|^4 + \frac{g_0}{3}|\Psi|^6. \qquad (2.61)$$

Since for small $|\Psi|^2$ in equilibrium [see (2.53)] $|\Psi_0|^2 = \alpha/\beta = a_0\tau^{2/3}/b_0$, this result is in agreement with (2.56), (2.57). Expression (2.61) is naturally so chosen as to correspond to the experiment. Parenthetically, the same method in application to the Ψ-theory of superconductivity was employed in paper [67], only not near but far from T_c. As far as I know, (2.61) was first applied by Y.G. Mamaladze [117]. Some other authors also discussed a generalization of the phase transition theory in the spirit of involving an equation of the type (2.61) (see references in [108]). Sobyanin and I developed the generalized Ψ-theory of superfluidity [108–112] on the basis of (2.61) which in turn underlay the 'generalized' Ψ-theory of superconductivity (see [90] and Sect. 2.3). But while the latter is of limited significance, the generalized Ψ-theory of superfluidity is a unique scheme capable of describing the behavior of liquid helium near the λ-point, not counting the incomparably more sophisticated approach based on the renormalization group theory (see [118] and references therein). In addition, this approach [118] is either of no or limited validity for the inhomogeneous and non-stationary cases.

Without going into details, we shall immediately present the expression for the involved free-energy density in some reduced units (instead of free energy, other thermodynamic potentials were used in [108–112] but this is of no importance):

$$F_{\mathrm{II}} = F_{\mathrm{I}} + \frac{3\Delta C_p}{(3+M)T_\lambda} \tag{2.62}$$

$$\times \left[-t|t|^{1/3}|\Psi|^2 + \frac{(1-M)|t|^{2/3}}{2}|\Psi|^4 + \frac{M}{3}|\Psi|^6 + \frac{\hbar^2}{2m}|\nabla\Psi|^2 \right].$$

Here $t = T_\lambda - T$, ΔC_p is the jump of specific heat determined by (2.53), M is the constant introduced in the theory, $\Psi = \Psi/\Psi_{00}$, $\Psi_{00} = \sqrt{1.43\rho_\lambda/m}$, $\rho_s = 1.43\rho_\lambda(T_\lambda - T)^{2/3}$. In the simplest version of the theory, we have $M = 0$ and, irrespective of this fact, the reduced order parameter Ψ is sometimes (for instance, in the vicinity of the axis of a vortex line) rather small and the term $|\Psi|^6$ in (2.63) can be ignored. A comparison with an experiment for Helium II leads to the estimate $M = 0.5 \pm 0.3$ (see [112]). The transition is second order for $M < 1$ and first order for $M > 1$.

For a shift of the λ-transition temperature in a film (for $M < 1$), we have

$$\Delta T_\lambda = T_\lambda - T_\lambda(d) = 2.53 \times 10^{-11} \left(\frac{3+M}{3} \right) d^{-3/2}\,(\mathrm{K}) \tag{2.63}$$

which generalizes (2.59) and corresponds to experimental data, and for a capillary with diameter d, the coefficient 2.53 in (2.63) is replaced by 4.76. Expressions for a number of other quantities (density, specific heat, etc.) are obtained and the effect of the external (gravitational, electric) fields, as well as Van der Waals forces, are taken into account. The behavior of ions in Helium II, the dependence of the density ρ_s on velocity v_s, and the vortex line structure are considered [119]. Furthermore, the theory is extended to the case of the presence of a flow of the normal part of a liquid (density ρ_n, velocity v_n) and the presence of dissipation and relaxation (for a non-stationary flow; for the initial Ψ-theory, this was done partially in [120]). The problem of vortex creation in a superfluid liquid (see [110] where the corresponding literature is cited) is very interesting. We note that, somewhat unexpectedly, this question proved to be of interest for simulating the process of creation of so-called topological defects in cosmology [121]. I believe that in an analysis of corresponding experiments the Ψ-theory of superfluidity may turn out to provide quite suitable methods.

The generalized Ψ-theory of superfluidity was not developed 'from first principles' or on the basis of a certain reliable microtheory (as in the situation with the Ψ-theory of superconductivity). This is a phenomenological theory that rests on the general theory of second-order phase transitions (Landau theory and scaling theory) and on experimental data [110, 111]. Such data is unfortunately quite insufficient for drawing a vivid conclusion concerning the region of applicability of the Ψ-theory. In the papers [122, 123], we find rather pessimistic judgements in this respect but Sobyanin was of the opinion that such a criticism is groundless. I do not hold any particular viewpoint here but my intuition suggests a very positive role of both the initial [96] and the generalized [108–118] Ψ-theories of superfluidity. In any case, clarification of the precision and the role of the Ψ-theory of superfluidity is currently pressing because experimental studies of superfluidity in Helium II are continued (see, for example, [124, 125]; see also comments 4*, 9*).

2.5 Thermoelectric Phenomena in Superconductors

Different papers have their own fate. My first paper [19] on superconductivity now seems dull to me and this is all from a bygone time. And, what concerns the second paper [22] accomplished in the same year, 1943, remains topical up to the present date. It was devoted to thermoelectric phenomena in superconductors. Before that, thermoelectric effects had been considered (see, for example, [58, 126]) to disappear completely in the superconducting state. Specifically, when a superconducting current passes through a seal of two superconductors, the Peltier effect is absent, the same as a noticeable thermoelectric current is absent upon heating one of the seals of a circuit consisting of two superconductors. But as a matter of fact, thermoelectric phenomena in superconductors do not vanish completely, although they can manifest themselves only under special conditions [22, 24]. The point is that in a superconductor one should take into account the possibility of the appearance of two currents – superconducting (the density $\boldsymbol{j}_\mathrm{s}$) and normal (the density $\boldsymbol{j}_\mathrm{n}$). In a non-superconducting (normal) state in a metal, there may flow only one current $\boldsymbol{j}$, Ohm's law $\boldsymbol{j} = \sigma \boldsymbol{E}$ holding in the simplest case. If there exists a gradient of chemical potential μ of electrons in a metal and a temperature gradient, then

$$\boldsymbol{j} = \sigma \left(\boldsymbol{E} - \frac{\nabla \mu}{e_0} \right) + b \nabla T. \tag{2.64}$$

In the superconducting state, as can readily be seen (see, for example, [128]), for the normal current, we have

$$\boldsymbol{j}_\mathrm{n} = \sigma_\mathrm{n} \left(\boldsymbol{E} - \frac{\nabla \mu}{e_0} \right) + b_\mathrm{n} \nabla T \tag{2.65}$$

instead of (2.3), and in the Londons' approximation (2.4) is preserved; instead of (2.5), we obtain

$$\frac{\partial (\Lambda \boldsymbol{j}_\mathrm{s})}{\partial t} = \boldsymbol{E} - \frac{\nabla \mu}{e_0} + \nabla \frac{\Lambda j_\mathrm{s}^2}{2 \rho_e} \tag{2.66}$$

where μ is the chemical potential of electrons and $\rho_e = e_0 n_\mathrm{s}$, n_s is the concentration of 'superconducting electrons' ($\boldsymbol{j}_\mathrm{s} = e_0 n_\mathrm{s} \boldsymbol{v}_\mathrm{s}$). Here, we omit the detail connected with the necessity of introducing different chemical potentials μ_n and μ_s in non-equilibrium conditions for a normal and superconducting electron subsystems (see [128]). Note that the last term on the right-hand side of (2.66) is of a hydrodynamic character [see (2.6)] and, in (2.5), it was omitted because of its small magnitude. However, the contribution of this term can be observed experimentally (see [128, 243, 246, 270–272] and references therein). Forgetting again about the last term in (2.66) in the stationary case for a superconductor, we have

$$E - \frac{\nabla \mu}{e_0} = 0 \tag{2.67}$$

from which it follows that [see (2.65)]:

$$j_n = b_n(T)\nabla T. \tag{2.68}$$

Thus, in a superconductor, the thermoelectric current j_n does not vanish completely. Why then is it not observed? As has already been mentioned, under particularly simple conditions, a normal current is totally compensated for by a superconducting current, i.e.,

$$j = j_s + j_n = 0, \quad j_s = -j_n. \tag{2.69}$$

By 'particularly simple conditions', we understand a homogeneous and isotropic superconductor, say, a non-closed small cylinder (a wire) on one end of which the temperature is T_1 and on the other end T_2 (we assume that $T_{1,2}$ is less than T_c).[9] In such a specimen, in the normal state (for $T_{1,2} > T_c$), we certainly have $j = 0$ and $E = \nabla \mu/e_0 - b\nabla T/\sigma$ [see (2.64)]; in the superconducting state, we of course also have $j = 0$ but [see (2.68) and (2.69)]:

$$j_s = -j_n = -b_n \nabla T, \quad E - \frac{\nabla \mu}{e_0} = 0. \tag{2.70}$$

If a superconductor is inhomogeneous and (or) anisotropic, then, generally speaking, the total compensation (2.69) does not occur and a certain, although weak, thermoelectric current must be [22] and is, in fact, observed [128, 129, 213]. But, one should not think that in the simple case considered earlier, when $j = 0$, all thermoelectric effects disappear. Indeed, the thermoelectric current j_n must be associated with some heat transfer, i.e., in superconductors, there must occur an additional (say, circulational or convective) heat transfer mechanism similar to the one that exists in a superfluid liquid.[10] This analogy was, properly speaking, the starting point for me in paper [22]. However, in [22], I made no estimate of the additional (circulational) thermal conductivity. Later I decomposed [64] the total heat conductivity $\varkappa$ involved into the heat transfer equation $q = -\varkappa \nabla T$ (q is the heat flux) into three parts: $\varkappa = \varkappa_{ph} + \varkappa_e + \varkappa_c$. Here $\varkappa_{ph}$ stands for the contribution due to phonons (the

---

[9] I did not want to place figures in this paper, although perhaps they would not be out of place here. But all the necessary illustrations concerning thermoeffects can be found in the readily available papers [128, 129, 213] and also in my Nobel Lecture [264], which is also published in this book.

[10] Such heat transfer is also possible in semiconductors that possess the corresponding electron and hole conductivities simultaneously (see [131]).

lattice), $\varkappa_e$ is due to electron motion such that there is no circulation (i.e., under the condition $j_n = 0$), and $\varkappa_c$ is due to circulation (convection). The estimates done in [64] indicated that $\varkappa_c$ must be negligibly small compared to $\varkappa_e$ but now, unfortunately, I do not understand these estimates.

After the BCS theory was created, it became possible to carry out a microscopic evaluation of $\varkappa_e$ and $\varkappa_c$. According to [130], at $T \sim T_c$,

$$\frac{\varkappa_c}{\varkappa_e} \sim \frac{k_B T_c}{E_F} \tag{2.71}$$

where E_F is the Fermi energy of electrons in a given metal.

This estimation was obtained earlier [222] on the basis of the two-fluid model and some assumptions. Finally, I also came to result (2.71) by estimating the thermal flux (heat transfer) due to creation of Cooper pairs in the colder end of a sample and their decay in the hot end [129, 132].[11] I had some doubts of whether the heat flux calculated on the basis of the kinetic equation [130] and an allowance of the effect on the boundaries [129, 132] should be summed up. Such an assumption is, however, erroneous: whenever the kinetic equation holds (i.e., the free path of 'normal electrons' is small compared to the sample length), the kinetic calculation, and the allowance for pair creation and breakdown on the boundaries are equivalent. However, the doubts in the validity of estimate (2.71) appeared to be useful since a more consistent estimation gave another result [213]:

$$\frac{\varkappa_c}{\varkappa_e} \sim \left(\frac{k_B T_c}{E_F} \right)^2 . \tag{2.72}$$

Apparently, kinetic calculations in [130] contained an error. The previously mentioned referred to isotropic superconductors but, in this case, the Seebeck coefficient $S = b/\sigma$ is known to be underestimated for the well-known reason by a quantity of the order of $E_F/k_B T$ (see [133, 134, 223]). Hence, for anisotropic and unconventional superconductors, estimate (2.71) is likely to be reasonable. For conventional isotropic superconductors at $T_c \sim (1\text{--}10)\,\mathrm{K}$ and $E_F \sim (3\text{--}10)\,\mathrm{eV}$, the convective thermal conductivity is quite negligible according to (2.72) because $\varkappa_c/\varkappa_e \lesssim 10^{-7}$. But for high-temperature superconductors at $T_c \sim 100\,\mathrm{K}$ and $E_F \sim 0.1\,\mathrm{eV}$, we already have $\varkappa_c/\varkappa_e \sim 0.1$ according to (2.71). The roughness of the estimate allows the suggestion that, in some cases, convective thermal conductivity may be appreciable. Therefore, I tried to explain [132] in this way the observed peak of thermal conductivity coefficient in HTSCs at $T \sim T_c/2$ (see [135–138]). However, this effect can also be explained by the corresponding temperature dependence of the coefficients $\varkappa_{ph}$ and $\varkappa_c$. This issue was discussed in the literature. Observation of the Righi–Leduc effect, also referred to as the thermal Hall effect [224], led to the

---

[11] This result was also presented in Sect. 5 of the paper *Phys.–Usp.* **40**, 407 (1997) before its modification.

conclusion [225] that the phonon part of thermal conductivity makes no case here (i.e., the contribution of the coefficient $\varkappa_{\mathrm{ph}}$ is insignificant; see [138,224]). At the same time, it is impossible to separate directly the contributions from $\varkappa_e$ and $\varkappa_c$ and I am not aware whether it can generally be done (an analysis is needed that would involve the role of anisotropy and the external magnetic field; see [213]).[12]

I have dwelt on the convective thermal conductivity (heat transfer) in superconductors at such length because I feel somewhat particularly unsatisfied in this respect. I have never properly investigated the microtheory or, as it is more often called, the electron theory of metals. That is why I was unable, and never even tried, to construct a consistent microtheory of convective heat transfer. Now it is certainly too late for me. But I hope that someone will investigate this problem sooner or later.

If a superconductor is not homogeneous and isotropic, as has already been mentioned, no complete compensation of currents $\boldsymbol{j}_{\mathrm{n}}$ and $\boldsymbol{j}_{\mathrm{s}}$ occurs, and some thermal currents must generally flow. The simplest cases are as follows: an isotropic but inhomogeneous superconductor and a homogeneous but anisotropic superconductor (monocrystal). More than 50 years ago (!), when paper [22] was written, alloys and generally inhomogeneous superconductors were thought of as something 'polluted', and it was not even clear whether the Londons' equations can be used in these conditions. For this reason, the case of inhomogeneous superconductors was only slightly touched upon in [22]. Concretely, it was pointed out that if in a bimetallic plate (say, different superconductors sealed or welded to each other), there is a temperature gradient perpendicular to the seal plane, an uncompensated current $\boldsymbol{j}$ is excited along the seal line, which runs around the seal: this generates a magnetic field perpendicular to both the plate and the seal line (see Fig. 3a in [128] and Fig. 3 in [129]). As I have said, such a version does not seem interesting. Attention was, therefore, given to a monocrystal with non-cubic symmetry when the tensor Λ_{ik} does not degenerate into a scalar (for cubic and isotropic superconductors $\Lambda_{ik} = \Lambda\delta_{ik}$). If, in such a plate-like crystal, the temperature gradient ∇T is not directed along the symmetry axis, a current $\boldsymbol{j}$ flowing round the plate is excited and a magnetic field H_{T}, proportional to $|\nabla T|^2$, is generated perpendicular to the plate. This field can easily be measured using modern methods. For details, see [22,128,129,140]. Unfortunately, attempts to observe the thermoelectric effect in question were made only in [141], the results of which remain ambiguous [128,140].

As it turned out, the thermoeffect for inhomogeneous isotropic superconductors is easier to analyze and easier to observe. For this purpose, it is most convenient to consider not a bimetallic plate but rather a superconducting ring (a circuit) consisting of two superconductors (with one seal at a temperature T_2 and the other at a temperature $T_1 < T_2$; see Fig. 3b in [128], Fig. 7 in [129], or Fig. 3 in [213]; see also my Nobel Lecture in this book). The pertinence of

---

[12] See also recent papers [244,245] devoted to heat transfer and related problems.

the choice of this particular version was indicated in [142, 143]. Paper [142] argued that this effect was quite different from that considered in [22] but this was a misunderstanding [128, 144]. Indeed, a bimetallic plate and a circuit of two superconductors differ topologically because of the presence of a hole in the latter case, which leads to the possibility of the appearance of a quantized magnetic-field flux through the hole (see Fig. 3 in [128]). A simple calculation (see [128, 129, 142–145]) shows that the flux through the indicated hole is equal to

$$\Phi = k\Phi_0 + \Phi_T, \quad \Phi_T = \frac{4\pi}{c} \int_{T_1}^{T_2} (b_{\mathrm{n,\,II}}\, \delta_{\mathrm{II}}^2 - b_{\mathrm{n,\,I}}\, \delta_{\mathrm{I}}^2)\, \mathrm{d}T$$

$$\Phi_0 = \frac{hc}{2e_0} = 2 \times 10^{-7}\ \mathrm{G\ cm^2}, \quad k = 0, 1, 2, \ldots \tag{2.73}$$

where the indices I and II refer to metals I and II forming the superconducting circuit, respectively, and $\delta \equiv \delta_0$ is the penetration depth: for $k = 0$, we obtain the result for a bimetallic plate. If we assume for simplicity that $(b_\mathrm{n}\delta^2)_\mathrm{II} \gg (b_\mathrm{n}\delta^2)_\mathrm{I}$ and $\delta_\mathrm{II}^2 = \delta_\mathrm{II}^2(0)(1-T/T_{\mathrm{c,\,II}})^{-1}$, then from (2.73), we obtain $(T_\mathrm{c} = T_{\mathrm{c,\,II}})$

$$\Phi_T = \frac{4\pi}{c}\, b_{\mathrm{n,\,II}}\delta_\mathrm{II}^2(0)\, T_\mathrm{c} \ln\left(\frac{T_\mathrm{c} - T_1}{T_\mathrm{c} - T_2}\right). \tag{2.74}$$

Estimates for tin $(b_\mathrm{n}(T_\mathrm{c}) \sim 10^{11}$–$10^{12}\mathrm{CGSE}$, $\delta(0) \approx 2.5 \times 10^{-6}\mathrm{cm})$ when $(T_\mathrm{c} - T_2) \sim 10^{-2}\mathrm{K}$, $(T_\mathrm{c} - T_1) \sim 0.1\,\mathrm{K}$, and generally $\ln[(T_\mathrm{c} - T_1)/(T_\mathrm{c} - T_2)] \sim 1$ lead to the value $\Phi_T \sim 10^{-2}\Phi_0$. Such a flux can readily be measured, and this was done in a number of papers as far back as 20 years ago (for the references, see [128, 145]). Here I will only refer explicitly to [146], which also confirmed the result (2.74).

As far as the thermoelectric current in a superconducting circuit is concerned, everything seems to be clear in principle, but this is not so. The point is that for a sufficiently massive and closed toroidal-type circuit (a hollow cylinder made of two superconductors), the measured flux $\Phi(T)$ appeared [145] to be several orders of magnitude higher than the flux (2.74) and, moreover, to possess a different temperature dependence. The origin of such an 'enormous' thermoeffect in superconductors has not yet been clarified. A probable explanation was suggested by R.M. Arutyunyan and G.F. Zharkov [147], although it has not yet been confirmed by an experiment. There are other explanations [246] of the results obtained in [145].[13] In this case, the measured flux through the hole is equal to $\Phi_T + k\Phi_0$ rather than Φ_T. As the critical temperature of the hottest seal T_2 approaches the temperature T_c of one of the

---

[13] In the recent paper [271] it is stated that the abovementioned ambiguity in the problem of thermocurrent is clarified if the so-called Bernoulli term $\nabla(\Lambda j_\mathrm{s}^2/2\rho_\mathrm{e})$ is taken into account in (2.66). One may hope that this issue will be made clear in the near future. The other new publications devoted to thermal effects are [270, 272].

superconductors, the resulting increase in thermoelectric current increases the entrapped flux $k\Phi_0$, i.e., a growth of k, energetically advantageous. This question was discussed in a number of papers [?, ?, 149, 150] but the mechanism responsible for the increase in the flux $\Phi(T)$ still remained unclear, and no new experiments have been carried out. The mechanism of vortex formation in the walls of a superconducting cylinder that leads to an increase of an entrapped flux with increasing thermoelectric current has been proposed only recently in [151].

It should be noted that (2.73), which also implies (2.74), is derived on the assumption that the total current density $j = j_s + j_n$ is zero throughout the entire circuit thickness. Meanwhile, near T_c, when the field penetration depth δ increases (more than this, $\delta \to \infty$ as $T \to T_c$), the current density j tends to the value corresponding to the current in the normal state (i.e., at $T > T_c$). Clearly, the flux Φ must then increase. Under such conditions, allowance should be made for the appearance in a superconductor of some charges (the so-called charge imbalance effect; see [145, 226] and other literature cited in [145, 213, 226]). It is only an allowance for the role of these charges that provides continuity for the transition from the normal to the superconducting state. By the way, near T_c, particularly when the coherent length ξ is small, the fluctuation effects also deserve attention. The influence of the charge imbalance effect upon the temperature dependence of the flux Φ in a superconducting ring was discussed in [227], where the effect was found to be small but the physical meaning of this result is not clear to me. I believe, in particular, that the allowance for flux entrapment (i.e., an increase in the number of trapped quanta of the flux with temperature) should be analyzed simultaneously with the allowance for the charge imbalance effect. The latter would also provide a clear insight (which in my opinion has not yet been attained) into the character and the results of measurements of thermal e.m.f. in the circuit upon a superconducting transition of one and then both of its units.

I turned [129, 132, 139] to the convective mechanism of thermal conductivity in superconductors many times and could not then understand why this issue was being ignored. Now (after paper [213]), the most probable explanation seems to be the fact that, within a correct kinetic calculation, the convective mechanism is involved automatically. Therefore, the contributions of $\varkappa_e$ and $\varkappa_c$ need not be separated from the observed coefficient of the electron component of thermal conductivity $\varkappa_e^{\mathrm{tot}} = \varkappa_e + \varkappa_c$. But is it always (when anisotropy and the action of external forces are involved) impossible to separate $\varkappa_e$ and $\varkappa_c$? This remains unclear to me. Furthermore, the coefficient $\varkappa_e$ can perhaps be determined by measuring the conductivity σ_n according to the Wiedemann–Franz law. Then, the coefficient $\varkappa_c$ will be determined as the difference $\varkappa_e^{\mathrm{tot}} - \varkappa_e$.

I hope, although not very much, that thermoeffects in superconductors (in a superconducting state) will no longer be ignored and there will finally appear corresponding experiments involving, in particular, HTSCs. In my opinion, it is nevertheless conceivable that the convective heat conduction mechanism plays a part in some cases.[11*]

Concluding this section, I would like to emphasize that, in accordance with the general context of this paper, I have only concentrated on those thermoelectric phenomena in superconductors which I investigated myself. Nevertheless, there exist some other related aspects of the problem. In this respect, I shall restrict myself by referring the reader to reviews [128,129,145] and the references therein, as well as to the books [40, 228] and the papers [152–154, 225, 270–272].[14*]

2.6 Miscellanea
(Superfluidity, Astrophysics, and Other Things)

As mentioned in Sect. 2.1, my first work [23] in the field of low-temperature physics, which was accomplished at the beginning of 1943, was devoted to light scattering in Helium II. This question was rather topical at that time because, when comparing the transition in helium and the Bose–Einstein gas condensation, one might expect very strong scattering near the λ-point. At the same time, the Landau theory [4] suggested no anomaly. But this was, so to say, a trivial result. The most interesting thing is that the scattering spectrum must consist not of the central line and a Mandelstam–Brillouin doublet, as in usual liquids, but of two doublets. Indeed, the Mandelstam–Brillouin doublet is associated with scattering on sound (or, more precisely, hypersonic) waves, while the central line is associated with scattering on entropy (isobaric) fluctuations. In the case of Helium II, and generally superfluid liquids, entropy fluctuations propagate (or, more precisely, dissipate) in the form of a second sound. This is the reason why, instead of a central peak, a doublet must be observed that corresponds to scattering on second sound waves also. In paper [23], I noted, however, that 'the inner anomalous doublet cannot be actually observed because on the one hand the corresponding splitting is too small $(\Delta\omega_2/\omega_2 \sim u_2/c \lesssim 10^{-7})$ and on the other hand, and this is particularly important, the intensity of this doublet relative to Mandelstam–Brillouin doublet is quite moderate'. Indeed, the inner-to-outer doublet intensity ratio is $I_2/I_1 \approx C_p/C_V - 1$ ($C_{p,V}$ is the specific heat at a constant pressure or for a constant volume). Even near the λ point, at low pressure in Helium II, we have $C_p/C_V = 1.008$. However, as in many other cases in physics, the pessimistic prediction did not prove to be correct. Firstly, the intensity of the inner doublet increases greatly with pressure and, secondly, and this is especially significant, the use of lasers promoted great progress in the study of light scattering. As a result, the inner doublet could be observed and investigated (see [155] and review [156], p. 830).

I have already mentioned papers [49, 106] devoted to superfluidity, to say nothing of papers [96, 108–112, 119] on the Ψ-theory of superfluidity. I would like also to mention the notes [157,158] whose titles cast light on their contents.

Finally, I shall dwell on the thermomechanical circulation effect in a superfluid liquid [144, 159]. In a ring-shaped vessel filled with a superfluid liquid (concretely, Helium II was discussed) and having two 'weak links' (for example, narrow capillaries), in the presence of a temperature gradient, there must occur a superfluid flow spreading to the entire vessel. Curiously, the conclusion concerning the existence of such an effect was suggested [144] by analogy with the thermoelectric effect in a superconducting circuit. At the same time, the conclusion was drawn concerning the existence of thermoelectric effects in superconductors [22], in turn, by analogy with the 'inner convection' occurring in Helium II in the presence of temperature gradient.

The effect under discussion was observed [160] but the accuracy of measurements of the velocity v_s was not enough to fix the jumps of circulation in superfluid helium (the circulation quantum is $2\pi\hbar/m_{He} \approx 10^{-3}\mathrm{cm}^2\mathrm{s}^{-1}$) which had been predicted by the theory [159]. Meanwhile, there exist interesting possibilities of observing not only jumps of circulation of a superfluid flow but also peculiar quantum interference phenomena (to this end, 'Josephson contacts' must be present in the 'circuit,' for example, narrow-slit diaphragms). In my opinion, the circulation effect in a non-uniformly heated ring-shaped vessel is fairly interesting, and not only for ^{4}He or solutions of ^{4}He with ^{3}He, but perhaps also in the case of the superfluidity of pure ^{3}He. Considering an extensive front of research in the field of superfluidity all over the world, I cannot understand why this effect is totally neglected. I do not know whether this is a matter of fashion, a lack of information, or something else.[14]

To save space in the other sections of the present paper, I shall mention here the works [114, 161–163]. The first of them [114] stresses the fairly obvious fact that molecular hydrogen H_2 does not become superfluid only for the reason that, at a temperature T_m exceeding the λ-transition temperature T_λ, it solidifies. As is well known, for H_2 the temperature T_m is $14\,\mathrm{K}$, whereas by estimation T_λ should be nearly $6\,\mathrm{K}$. Perhaps liquid hydrogen may be supercooled, for example, by way of expansion (a negative pressure), application of some fields and the use of films on different substrates as well as in the dynamical regime.

The possibility of observing the secondary sound and convective heat transfer in superconductors, in the first place accounting for exciton-type excitations (we mean bosons) was considered in [161]. I should say that paper [161] was written in 1961 and I am unaware of the present state of the questions discussed in it.

In 1978, there appeared reports on the observation of a very strong diamagnetism (superdiamagnetism) in CuCl, when the magnetic susceptibility χ is negative, and $|\chi| \sim 1/4\pi$ (of course, $|\chi| < 1/4\pi$ because $\chi = -1/4\pi$ corresponds to an ideal diamagnetism). After that (in 1980), there appeared

---

[14] A.A. Sobyanin has pointed out the interesting possibility of 'spinning-up' the normal component of Helium II inside a vessel by means of electric and magnetic fields acting on the helium ions.[3*]

indications of the existence of superdiamagnetism in CdS, too. What it was that was actually observed in the corresponding experiments (for references see [162]) remains still unclear and this question was somehow 'drawn in the sand'.[5*] Many physicists believe that the measurements were merely erroneous. In any case, attempts were made to associate the observations with the possibility of the existence of superdiamagnetics other than superconductors.[15]

The last study in this direction in which I took part was reported in [162]. Further on, the question of superdiamagnetism somehow 'faded away' (see, however, [164]) and I am unacquainted with the progress in this field. When seeking ways of explaining superdiamagnetism, I made an attempt to generalize the Ψ-theory of superconductivity [163]. It is unknown to me whether this paper is of any value now.

Concluding this section, I shall dwell on an astrophysical problem, namely the possibility of the existence of superconductivity and superfluidity in space. It seems to me that a small digression will not lead us beyond the scope of the general context of the paper. When I was young and then middle-aged, I used to entertain myself by doing an exercise which I called then a brainstorming (I wrote about it in my book [1], p. 305).[6*] The procedure of the 'attack' was as follows: looking at my watch, I set myself a task to think up some effect within a certain time interval, say, within 15–30 min. Here is a concrete example. If I am not mistaken, it was 1962, and I was travelling by train from Kislovodsk to Moscow. I was alone in the compartment with no book to read and so decided to conceive of something. I had been engaged in low-temperature physics and astrophysics for a number of years and, therefore, a natural question for me was where and under what conditions superfluidity and superconductivity could be observed in space. To formulate a question is frequently equivalent to doing half the work. It actually took me no more than the prescribed time to think that the existence of superfluidity is possible in neutron stars and superconductivity in the atmosphere of white dwarfs and that there may exist superfluidity of the neutrino 'sea.' On returning to Moscow, I took up all three problems – the first two together with D.A. Kirzhnits [165, 166] and the third in collaboration with G.F. Zharkov [167].

The interaction between neutrons with antiparallel spins in the s-state corresponds to attraction and, therefore, in a degenerate neutron gas, there will appear pairing in the spirit of BCS theory. For the gap width $\Delta(0) \sim k_B T_c$, we obtained the estimate $\Delta(0) \sim (1\text{–}20)$ MeV, i.e., in the center of a neutron star (for a density $\rho \sim 10^{14}\text{–}10^{15} \mathrm{g\,cm^{-3}}$) we obtained $T_c \sim 10^{10}\text{–}10^{11}$K, while, on the neutron phase boundary (for $\rho \sim 10^{11} \mathrm{g\,cm^{-3}}$), we had $T_c \sim 10^7$K. It

---

[15] In these experiments, a very strong diamagnetism was observed but the conductivity of the samples was not at all anomalously large. Such a situation is also possible for superconductors in the case of superconducting seeds (granules) which are separated by non-superconducting layers. The question, however, arose as to whether or not superdiamagnetism can be observed in dielectrics and non-superconductors in general.

was also indicated that the rotation of a neutron star results in the formation of vortex lines. The fact that, in nuclear matter, superfluidity may occur had actually been known before but applied to neutron stars (at that time, in 1964, they had not yet been discovered), as far as I know, our paper was pioneering. Incidentally, in [168], where I summarized my activity in the field of superfluidity and superconductivity in space, I also pointed to a possible superconductivity of nuclei-bosons (for example, α-particles) in the interior of white dwarfs and to the superconductivity of protons which are present in a certain amount in neutron stars.

The possibility of the existence of superconductivity in some surface layer of the cold stars–white dwarfs was discussed in papers [166,168]. The estimates give little hope. For example, for a density $\rho \sim 1\,\mathrm{g\,cm^{-3}}$ the temperature is $T_c \sim 200\,\mathrm{K}$ and, as the density increases, T_c falls rapidly. Somewhat more interesting is the possibility of the superconductivity of metallic hydrogen in the depths of large planets – Jupiter and Saturn [168]. The estimates of the critical temperature T_c for metallic hydrogen, which are known from the literature, reach 100–300 K but the temperature in the depth of the planets is unknown. I am unacquainted with the present-day state of the problem but it seems to me that the existence of superconductivity in stars and large planets is hardly probable. The possibility of the appearance of superfluidity in the degenerate neutrino 'sea,' whose existence at the early stages of cosmological evolution was discussed in some papers, was considered in note [167] (see also [168]). Such a possibility, as applied to neutrinos or some hypothetical particles now involved in the astrophysical arsenal, is currently of no particular interest, but nevertheless it is reasonable to bear in mind.

2.7 High-Temperature Superconductivity

Beginning in 1964, I started investigating high-temperature superconductivity (HTSC) and from that time this problem remained, and remains, at the center of my attention, although I was interested in many other things as well. My story about this work should, however, begin with quite a different question that concerns surface superconductivity. This question is as follows: can two-dimensional superconductors in which the electrons (or holes) participating in superconductivity are concentrated near the boundary of, say, a metal or a dielectric with a vacuum, on the boundary between, e.g., twins (i.e., on the boundary of twinning), etc. exist? It seems to me that surface superconductivity might be particularly well-pronounced for electrons on surface levels which were first considered by I.E. Tamm as far back as 1932 [169]. The possibility of this particular superconductivity was discussed in paper [170]. The answer was affirmative – the Cooper pairing and the whole BCS scheme works in the two-dimensional case as well. The following possibility was also pointed out: electrons are located at volume-type levels but their attraction, which leads to superconductivity, takes place only near the body surface (or on the twinning boundary). Note that surface ordering, although absent in the volume, may certainly take place not only in the case of superconductivity; it is also possible, for example, for ferro- and antiferromagnetics, and ferroelectrics [171]. I subsequently saw experimental research testifying to the realistic character of such situations. But I did not follow the appearance of the corresponding literature and cannot, therefore, give any references. Besides, this is not the subject of the present paper. As to surface superconductivity, it was emphasized in 1967 that long-range superconducting order is impossible in two dimensions [172]. At the same time, as distinguished from the one-dimensional case, in two dimensions (the case of a surface) the fluctuations that destroy the order increase with the surface size L only logarithmically. Accordingly, even for surfaces of macroscopic size ($L \gg a$, where a is atomic size), the fluctuations may not be so large [173]. An even more important circumstance is that, in a two-dimensional system, there may occur a quasi-long-range order under which superfluidity and superconductivity are preserved. This is an extensive issue and I, therefore, restrict ourselves to mentioning paper [174] and the monograph [175] (Chap. 1, Sect. 5 and Chap. 6, Sect. 5), where one can find the corresponding citations.[6*] Briefly speaking, superconductivity may well exist in two-dimensional systems. From an electrodynamic point of view, surface superconductors must behave as very thin superconducting films [176, 177]. In a certain sense, surface superconductivity is realized. For instance, superconductivity is observed in a $NbSe_2$ film with a thickness of only two atomic layers [178]. It would be more interesting to obtain surface superconductors on the Tamm (surface) levels [170]. It is obvious how interesting, and probably important from the point of view of applications, would be a dielectric possessing surface superconductivity. I am not, however, definitely sure that such a version may be thought of as radically different from a dielectric covered itself by a superthin superconducting film. But, after all, the difference does exist. The problem of surface superconductivity seems to be demanding and significant, irrespective of the corresponding value of the critical temperature T_c.

The fates decreed, however, that surface superconductivity was to be associated with the problem of HTSC. To be more precise, the association appeared in my own work.

Before clarifying the matter, I shall make several remarks (henceforth, I shall sometimes use the text of my paper [179] which may prove to be unavailable to the reader).

For a full 65 years, the science of superconductivity was part of low-temperature physics, i.e., temperatures of liquid helium (and, in some cases, liquid hydrogen). Thus, for example, the critical temperature of the first known superconductor, mercury, discovered in 1911, is $T_c = 4.15\,K$, and the critical temperature of lead, whose superconductivity was discovered in 1913, is $T_c = 7.2\,K$. If I am not mistaken, higher T_c values were not achieved until 1930, although it was definitely understood that higher T_c were desirable. The next important step on this way was the synthesis of the compound

Nb_3Sn with $T_c = 18.1\,K$ in 1954. Despite a great effort, it was not until 1973 that the compound Nb_3Ge with $T_c = 23.2$–$24\,K$ was synthesized. Subsequent attempts to raise T_c were unsuccessful until 1986, which saw the first indications (soon confirmed) of superconductivity in the La–Ba–Cu–O system with $T_c \sim 35\,K$ [180]. Finally, in early 1987, a truly HTSC $YBa_2Cu_3O_{7-x}$ with $T_c = 80$–$90\,K$ was created [181]. (This statement reflects my opinion that the term 'high-temperature' is appropriate only for superconductors with $T_c > T_{b,\,N_2} = 77.4\,K$, where, obviously, $T_{b,\,N_2}$ is the boiling nitrogen temperature at atmospheric pressure.)

The discovery of HTSCs became a sensation and gave rise to a real boom. One of the indicators of this boom is the number of publications. For example, in the period of 1989–1991, about 15,000 papers devoted to HTSC appeared, i.e., on average, approximately 15 papers a day.[7*] For comparison, one of the reference books states that, in the 60 years from 1911 to 1970, about 7,000 papers in total were devoted to superconductivity. Another indicator is the scale of conferences devoted to HTSC. Thus, at the conference $M^2HTSC\ III$ in Kanazawa (Japan, July 1991) there were approximately 1,500 presentations and the conference proceedings occupied four volumes with a total size of over 2,700 pages (see [182]). Undoubtedly, such a scale of research is, to a large extent, explained by the high expectations for HTSC applications in technology. These expectations, by the way, from the very beginning, appeared to me to be somewhat exaggerated, and this was later confirmed in practice. But, of course, the potential importance of HTSC for technology, medicine (nuclear magnetic resonance tomograph), and physics itself leaves no doubts. Nevertheless, I still do not completely understand such a hyperactive reaction from the scientific community and the general public to the discovery of HTSC: it is some sort of social phenomenon.

Another phenomenon that may be attributed either to sociology or to psychology is the complete oblivion to which HTSC researchers, who began working successfully in 1986, consigned their predecessors. Indeed, the problem of HTSC was born not in 1986 but at least 22 years earlier – in its current form, this problem was first stated by W.A. Little in 1964 [183]. Firstly, Little posed the question: why was the critical temperature of the superconductors known at the time not so high? Secondly, he pointed out a possible way of raising T_c to the level of room temperature, or even higher. Specifically, Little proposed replacing the electron–phonon interaction, responsible for superconductivity in the Bardeen, Cooper, and Schrieffer (BCS) model [18], by the interaction of conduction electrons with bound electrons or, in a different terminology which Little did not use, with excitons. In terms of the well-known BCS formula for the critical temperature

$$T_c = \theta \exp\left(-\frac{1}{\lambda_{\text{eff}}}\right) \qquad (2.75)$$

the meaning of the exciton mechanism is that the region of attraction between conduction electrons θ is set to be $\theta \sim \theta_{\text{ex}}$, where $k_B\theta_{\text{ex}}$ is the characteristic exciton energy. In contrast, for the electron–phonon mechanism of attraction in (2.75), we have $\theta \sim \theta_D$, where θ_D is the Debye temperature of the metal. Since the situation in which $\theta_{ex} \gg \theta_D$ is quite possible and even typical, it follows that, for the same value of the effective dimensionless interaction parameter λ_{eff}, for the exciton mechanism T_c is θ_{ex}/θ_D times higher than for phonons. Concretely, Little proposed to create an 'excitonic superconductor' on the basis of organic compounds by designing a long conducting (metallic) organic molecule (a 'spine') surrounded by side 'polarizers' – other organic molecules [183].

It is not appropriate to go into details here. Let me just point out that Little's work did not remain unnoticed. Quite the opposite: it attracted a lot of attention. In particular, I also followed up Little's work by suggesting a somewhat different version: roughly speaking, replacing the quasi-one-dimensional conducting thread in Little's model with a quasi-two-dimensional structure ('sandwich'), i.e., with a conducting thin film placed between two 'polarizers' (dielectric plates) [184]. More precisely, in paper [184], with a reference to the paper [170] on surface superconductivity, it was assumed that T_c may be raised with the help of some dielectric coverings of metallic surfaces. It was emphasized that quasi-two-dimensional structures are much more advantageous than quasi-one-dimensional structures [183] because of the considerably smaller role of fluctuations (this argument was worked out in [173]). Later on, I became engaged in earnest in the HTSC problem and concentrated on 'sandwiches', i.e., thin metallic films in dielectric and semiconducting 'coatings' and on layered superconducting compounds – these kind of 'files' of sandwiches [175,185–189].

I should say that I write rather easily and, moreover, I even feel the necessity of expressing my thoughts in written form. As a result, during the 32 years in which I have been interested in the HTSC problem, I wrote many (probably, too many) papers on the subject, particularly popular papers. I do not think I need to refer to many of them here. Among the published works, special attention is deserved by the monograph [175]. This book was the outcome of the joint efforts undertaken by L.N. Bulaevskii, V.L. Ginzburg, D.I. Khomskii, D.A. Kirzhnits, Y.V. Kopaev, E.G. Maksimov, and G.F. Zharkov (the I.E. Tamm Department of Theoretical Physics of the P.N. Lebedev Physical Institute of the USSR Academy of Sciences, Moscow) who had been 'attacking' the HTSC problem for several years. This monograph was published in Russian in 1977 and in an English translation in 1982, and was the first and, up to 1987, the only one devoted to this issue. In [175], a whole spectrum of possible ways of obtaining HTSC was considered.

I shall now dwell on some of the results of our work.

A very important question is whether or not there are some limitations on admissible T_c values in metals, say, due to the requirement of crystal lattice stability. Such limitations are possible in principle and, moreover, in the 1972 paper [190], it was stated that it was the requirement of lattice stability that fully obstructs the possibility of the existence of HTSC. The point is that the dimensionless parameter of the interaction force λ_{eff} in the BCS formula

(2.75) can be written in the form

$$\lambda_{\text{eff}} = \lambda - \mu^* = \lambda - \frac{\mu}{1 + \mu \ln(\theta_{\text{F}}/\theta)}. \tag{2.76}$$

Here λ and μ are, respectively, the dimensionless coupling constants for phonon or exciton attraction and Coulomb repulsion, and $k_{\text{B}}\theta_{\text{F}} = E_{\text{F}}$ is the Fermi energy. At the same time, in the simplest approximation (homogeneity and isotropy of material, and weak coupling), we have

$$\mu - \lambda = \frac{4\pi e^2 N(0)}{q^2 \varepsilon(0, q)} \tag{2.77}$$

where $\varepsilon(\omega, q)$ is the longitudinal permittivity for the frequency ω and for the wavenumber q, and the factor $1/q^2\varepsilon(0,q)$ should be understood as a certain mean value in $\boldsymbol{q}$, and $N(0)$ is the density of states on the Fermi boundary for a metal in the normal state. If, as was assumed in [190], the stability condition has the form

$$\varepsilon(0, q) > 0 \tag{2.78}$$

then, from (2.77), it follows that,

$$\mu > \lambda. \tag{2.79}$$

Both this inequality and (2.76) imply that superconductivity (for which, certainly, $\lambda_{\text{eff}} > 0$) is generally possible only due to the difference between μ^* and μ, the T_c value being not large. It was, however, already known empirically that $\mu < 0.5$ and sometimes $\lambda > 1$ and, thus, that inequality (2.79) is violated. Apart from this and some other arguments already expressed in the early stages [188], it was later shown strictly (see [175, 191, 192] and the literature cited there) that the stability condition (2.78) is invalid and, in fact, the stability condition has the form (for $q \neq 0$)

$$\frac{1}{\varepsilon(0, q)} \leq 1 \tag{2.80}$$

i.e., is satisfied if one of the inequalities

$$\varepsilon(0, q) \geq 1, \quad \varepsilon(0, q) < 0 \tag{2.81}$$

holds. It is interesting that the values $\varepsilon(0, q) < 0$ for large q, important in the theory of superconductivity, are realized in many metals [193, 194]. From the second inequality (2.81) and (2.77), it is obvious that the parameter λ may exceed μ. On the basis of this fact, our group came to the conclusion even before 1977 (I mean in the Russian edition of the book [175]) that the general requirement of stability does not restrict T_c and it is quite possible, for example, that $T_c \lesssim 300\,\text{K}$.

As has already been mentioned, the idea of the exciton mechanism is connected with the possibility of raising T_c by increasing the temperature θ

in (2.75) which determines the energy range $k_B\theta$ where the electrons attract one another near the Fermi surface and, thus, form pairs. It is assumed that weak coupling takes place here, when

$$\lambda_{\text{eff}} \ll 1. \tag{2.82}$$

It is only under this condition that (2.75) and the BCS model are applicable. But, the BCS theory is, on the whole, more extensive and admits consideration of the case of strong coupling [195], when

$$\lambda_{\text{eff}} \gtrsim 1. \tag{2.83}$$

Under conditions (2.83) for the strong coupling formula, (2.75) is, of course, already invalid although it is clear from it that the temperature T_c rises with increasing λ_{eff}. In the literature, a large number of expressions for T_c are proposed for the case of strong coupling (see [175, 192, 196, 197] and some references therein). The simplest of these expressions is as follows:

$$T_c = \theta \exp\left(-\frac{1+\lambda}{\lambda - \mu^*}\right). \tag{2.84}$$

Exactly as it should be under weak coupling conditions (2.82) or, more precisely, under the condition $\lambda \ll 1$, formula (2.84), of course, becomes (2.75). If, in (2.84), we set $\mu^* = 0.1$ then, for example, for $\lambda = 3$ the temperature is $T_c = 0.25\theta$. Therefore, for the value $\theta = \theta_D = 400\,\text{K}$, which is readily admissible for the phonon mechanism, we already have $T_c = 100\,\text{K}$. More accurate formulae also suggest that, for strong coupling (2.83), the phonon mechanism can already allow temperatures $T_c \sim 100\,\text{K}$ and even $T_c \sim 200\,\text{K}$. But, the analysis carried out in [175] and later showed that for 'conventional' superconductors with strong coupling, the temperature T_c is rather small. For example, for lead we have $\theta_D = 96\,\text{K}$ and, therefore, in spite of the high value $\lambda = 1.55$, the critical temperature is $T_c = 7.2\,\text{K}$. For such a conclusion, i.e., that θ_D falls with increasing λ, there also exist theoretical arguments (see [175], Chap. 4). That was the reason why we (or, at least, I) did not hope for the creation of HTSCs at the expense of strong coupling but possessing the phonon mechanism. In any case, as I have already mentioned, in [175], a versatile and unprejudiced approach to the HTSC problem prevailed. Here I cite the last part of Chap. 1 written by myself for the book [175]:

'On the basis of general theoretical considerations, we believe at present that the most reasonable estimate is $T_c \lesssim 300\,\text{K}$, this estimate being, of course, for materials and systems under more or less normal conditions (equilibrium or quasi-equilibrium metallic systems in the absence of pressure or under relatively low pressures, etc.). In this case, if we exclude from consideration metallic hydrogen and, perhaps, organic metals, as well as semimetals in states near the region of electronic phase transitions, then it is suggested that we should use the exciton mechanism of attraction between the conduction electrons.

In this scheme, the most promising materials from the point of view of the possibility of raising T_c, are apparently layered compounds and dielectric–metal–dielectric sandwiches. However, the state of the theory, let alone the experiment, is still far from being such as to allow us to regard as closed other possible directions, in particular, the use of filamentary compounds. Furthermore, for the present state of the problem of high-temperature superconductivity, the most sound and fruitful approach will be one that is not preconceived, in which attempts are made to move forward in the most diverse directions.

The investigation of the problem of high-temperature superconductivity is entering into the second decade of its history (if we are talking about the conscious search for materials with $T_c \gtrsim 90\,\mathrm{K}$ using exciton and other mechanisms). Supposedly, there begins at the same time a new phase of these investigations, which is characterized not only by greater scope and diversity but also by a significantly deeper understanding of the problems that arise. There is still no guarantee whatsoever that the efforts being made will lead to significant success, but a number of new superconducting materials have already been produced and are being investigated. Therefore, it is in any case difficult to doubt that further investigations of the problem of high-temperature superconductivity will yield many interesting results for physics and technology, even if materials that will remain superconducting at room (or even liquid-nitrogen) temperatures will not be produced. However, as has been emphasized, this ultimate aim does not seem to us to have been discredited in any way. As may be inferred, the next decade will be crucial for the problem of high-temperature superconductivity.'

This was written in 1976. Time passed, but the multiple attempts to find a reliable and reproducible way for creating HTSC have been unsuccessful. As a result, after the burst of activity came a slackening which gave cause for me to characterize the situation in a popular paper [198] published in 1984, as follows:

'It somehow happened that research into high-temperature superconductivity became unfashionable (there is good reason to speak of fashion in this context since fashion sometimes plays a significant part in research work and in the scientific community). It is hard to achieve anything by making admonitions. Typically, it is some obvious success (or reports of success, even if erroneous) that can radically and rapidly reverse attitudes. When they sense a 'rich strike,' the former doubters, and even dedicated critics, are capable of turning coat and become ardent supporters of the new work. But this subject belongs to the psychology and sociology of science and technology. In short, the search for high-temperature superconductivity can readily lead to unexpected results and discoveries, especially since the predictions of the existing theory are rather vague.'

I did not expect, of course, that this 'prediction' would come true in two years [180, 181]. It came true not only in the sense that HTSCs with $T_c > T_{\mathrm{b,N_2}} = 77.4\,\mathrm{K}$ were obtained but also, so to say, in the social aspect: as I have already mentioned, a real boom began and a 'HTSC psychosis' started. One of the manifestations of the boom and psychosis was an almost total oblivion to everything that had been done before 1986, as if discussion of the HTSC problem had not begun 22 years before [183,184]. I have already dwelt on this subject and in the papers [179,196] and would not like to return to it here. I will only note that J. Bardeen, whom I always respected, treated the HTSC problem with understanding both before 1986 and after it (see [199]).

The present situation in solid state theory and, in particular, the theory of superconductivity does not allow us to calculate the temperature T_c or indicate, with sufficient accuracy and certainly, especially in the case of compound materials, what particular compound should be investigated. Therefore, I am of the opinion that theoreticians could not have given experimenters better and more reliable advice as to how and where HTSC could be sought than was done in the book [175]. An exception is perhaps only an insufficient attention to the superconductivity of the $BaPb_{1-x}Bi_xO_3$ (BPBO) oxide discovered in 1974. When $x = 0.25$, for this oxide, we have $T_c = 13\,K$ which is a high value for a T_c when it is estimated in a way similar to that used for conventional superconductors. In the related oxide $Ba_{0.6}K_{0.4}BiO_3$ (BKBO), superconductivity with $T_c \sim 30\,K$ was discovered in 1988. Most importantly, the compound $La_{2-x}Ba_xCuO_4$ (LBCO) in which superconductivity with $T_c \sim 30\text{–}40\,K$ was discovered in 1986 [180] and is thought that the discovery of HTSC belongs to the oxides. However even now, 10 years later, one cannot predict, even roughly, the values of T_c for a particular material and, moreover, even the very mechanism of superconductivity in cuprates and, in particular, in the most thoroughly investigated cuprate $YBa_2Cu_3O_{7-x}$ (YBCO) with $T_c \sim 90\,K$ is not yet clear.

It is inappropriate to dwell here extensively on the current state of the HTSC problem. I shall restrict myself to several remarks. At first glance, HTSC cuprates differ strongly from 'conventional' superconductors (see, for example, [53,182,200,214]). This circumstance gave rise to the opinion that HTSC cuprates are something special – either the BCS theory is inapplicable to them or, in any case, a non-phonon mechanism of pairing acts in them. This tendency was very clearly expressed at the 1991 M^2HTSC III conference [182].

Indeed, the phonon mechanism has no exclusive rights. In principle, the exciton (electronic) mechanism, the Schafroth mechanism (creation of pairs at $T > T_c$ with a subsequent Bose–Einstein condensation), the spin mechanism (pairing due to exchange of spin waves or, as it is sometimes called, due to spin fluctuations), and some other mechanisms (for some more details and references see, for example, [197,214]) may all exist. Since I have always been a supporter of the exciton mechanism, I would be only glad if this very mechanism proves to act in HTSC. However, there is not yet any grounded basis for such a statement. In the BKBO oxide and in doped fullerenes (fullerites) of the K_3C_{60} and Rb_3C_{60} type (they all possess a cubic structure) with $T_c \sim 30\text{–}40\,K$, the phonon mechanism obviously prevails. The same relates to superconductivity in MgB_2 at $T_c = 40\,K$ [260,261] (see Chap. 6 in [2]).

The situation is more complicated with cuprate oxides, which also are highly anisotropic layered compounds. However, E.G. Maksimov, O.V. Dolgov, and their colleagues indicate, I believe, convincingly, that the phonon mechanism may quite possibly also dominate in HTSC cuprates. In any case, omitting the important question of a 'pseudogap' [233], HTSC cuprates in the normal state differ from ordinary metals in only a quantitative respect. Formally, a standard electron–phonon interaction with a coupling constant $\lambda \approx 2$ accounts well for the high values $T_c \sim 100$–$125\,\mathrm{K}$ as being due to the high Debye temperature $\theta_D \sim 600\,\mathrm{K}$ (see [197, 201–205, 256] and the literature cited there).[16] The properties of the superconducting state of HTSC cuprates are a more complicated entity. To explain them, it is already insufficient to use a standard isotropic approximation in the model of a strong electron–phonon interaction. However, allowing for the anisotropy of the electron spectra and interelectron interaction, the electron–phonon interaction all the same may play a decisive role in the formation of a superconducting state. As has been shown [215, 216] (see also [206–208]), in the framework of multi-zone models allowing for standard electron–phonon and Coulomb interactions, one can obtain a strongly anisotropic superconducting gap including its sign reversal in the Brillouin zone, which imitates d-pairing. It is also possible that the electron–exciton interaction and peculiarities of the electron spectrum, which are almost insignificant for understanding the properties of the normal state, make their contribution to the formation of the superconducting state. I do not regard myself competent enough to think of such statements as proved. But it is beyond doubt that a general denial of the important role of the phonon mechanism of HTSC (in cuprates) typical of the recent past (see [182]) is already behind us [204, 256, 267, 268] (possibly I only hope so, see [263]).[10*]

Suppose, for the sake of argument, that in the already known HTSCs the exciton mechanism does not play any role. This is, of course, important and interesting but in no way discredits the very possibility of a manifestation of the exciton mechanism. As has already been mentioned, we are not aware of any evidence contradicting the action of the exciton mechanism. But, it is actually not easy for the exciton mechanism to manifest itself. This will require some special conditions which are not yet clear (see, in particular, [205]).

The highest critical temperature fixed today (for $HgBa_2Ca_2Cu_3O_{8+x}$ under pressure) reaches $164\,\mathrm{K}$. Such a value can be attained with the phonon mechanism. But if one succeeds in reaching a temperature $T_c > 200\,\mathrm{K}$, the phonon mechanism will hardly be sufficient (when $\lambda = 2$, the temperature $T_c = 200\,\mathrm{K}$ is obtained for $\theta_D \approx 1000\,\mathrm{K}$). As to the exciton mechanism, even room temperature is not a limit for T_c. A search for HTSCs with the highest

---

[16] I find it necessary to note that the report [201] was, in fact, prepared by E.G. Maksimov alone. My name appeared in [201] only because there was a difficulty with including this report on the agenda and I had, by Maksimov's consent, to include my name which enabled him to participate in the 1994 M^2HTSC IV conference. It is not a pleasure to speak about such morals and manners, but this is the truth.

possible critical temperatures is now being and will, of course, be undertaken. It seems to me, as before, that the most promising in this respect are layered compounds and dielectric–metal–dielectric 'sandwiches.'[17] It would be natural to use the atomic layer-by-layer synthesis here [209, 218, 247]. The role of a dielectric in such sandwiches can be played by organic compounds in particular. Still and all, the possibilities that may open on the way are virtually boundless. It is, therefore, especially reasonable to be guided by some qualitative consideration (see, for example, [175], Chap. 1).

For 22 long years (from 1964 to 1986), which, however, flew by very quickly, HTSC was a dream for me and to think of it was something like a gamble. Now it is an extensive field of research, tens of thousand papers are devoted to it, and hundreds or even thousands of researchers are engaged in the study of one or another of its aspects. Much has already been done, but much remains to do. Even the mechanism of superconductivity in HTSC cuprates is rather obscure, to say nothing of the myriad particular questions. I think that among these questions the first place belongs to the question of the maximum attainable value of the critical temperature T_c under not very exotic conditions, say, at atmospheric pressure and for a stable material. More concretely, one can pose a question concerning the possibility of creating superconductors with T_c values lying within the range of room temperatures (the problem of room temperature superconductivity (RTSC)). RTSC is, in principle, possible but there is no guarantee in this respect. The problem of RTSC has generally taken the place that had been occupied by HTSC before 1986–1987. I am afraid that I do not see any possibility for myself to do something positive in this direction and it only remains to wait impatiently for coming events (see Chap. 3 in this book).

2.8 Concluding Remarks

By 1943, when I began studying the theory of superconductivity, 32 years had already passed since the discovery of the phenomenon. Nonetheless, at the microscopic level, superconductivity had not yet been understood and had actually been a 'white spot' in the theory of metals and, perhaps, in the whole physics of condensed media. The superfluidity of Helium II had been discovered in its explicit form no more than 5 years before that time, and its connection with superconductivity had only been outlined. The world was in a terrible war and I myself hardly understand now why the enigmas of low-temperature physics seemed so tempting to me when I was cold and semi-starving in evacuation in Kazan. But it was so. A poor command of mathematics, an inability to concentrate on one particular task (I was simultaneously engaged in several problems), and difficulties in the exchange of

---

[17] In addition to intuitive arguments [175, 186, 188, 189], there are also some concrete arguments [201, 205] in favor of such quasi-two-dimensional structures.

scientific information, especially with experimenters, in the war and post-war years obstructed a rapid advance, and it was only in 1950 that something appeared completed (I mean the Ψ-theory of superconductivity). But this completeness is, of course, rather conditional because new questions and problems constantly arose.

At the same time, the character of studies in the field of low-temperature physics, as well as the whole of physics, was changing radically. It is even hard to imagine now that it was only one laboratory that succeeded in obtaining liquid helium between 1908 and 1923. It is hard to imagine that applications of superconductivity in physics, to say nothing of technology, were fairly modest for three decades. And it was not until the 1960s that strong superconducting magnets were created and extensively used. At the present time, superconductivity finds numerous applications (see, for example, [71, 210]). Even the small book [211] intended for schoolchildren presents various applications of superconductivity, including giant superconducting magnets in tokamaks and tomographs. The creation of HTSCs (1986–1987) gave rise to great expectations of the possibility of new applications of superconductivity. These expectations were partly exaggerated but nevertheless now, after 20 years, much has already been done in this direction, even in respect of electric power lines and strong magnets [212], not to mention some other applications [219]. I wrote in Sect. 2.7 about the boom provoked by the creation of HTSC. Many thousands of papers and hundreds or even thousands of researchers – what a contrast with what was observed in, say, 1943 or as recently as 20 years ago!

In the light of the present state of the theory of superconductivity and superfluidity, much of what has been said in this paper is only of historical interest and, in other cases, is somewhere far from the forefront of the current research. At the same time, and this is very important, I have mentioned a large number of questions and problems which still remain unclear. This lack of clarity concerns the development of the Ψ-theory of superconductivity and its application to HTSC, the application of the Ψ-theory of superfluidity, the problem of surface (two-dimensional) superconductivity, the question of thermoeffects in superconductors (and especially their connection with heat transfer, see also [257]), the circulation effect in a non-uniformly heated vessel filled with a superfluid liquid, and some other things, to say nothing of HTSC theory (see [256] for a review) and also Chap. 3 in the present book and references therein); for further investigation of ferromagnet superconductors see, for example, [274]. The aim of the paper will have been attained if it at least helps to draw attention of both theoreticians and experimenters to these problems.[12*, 13*]

Acknowledgments

Taking the opportunity, I express my gratitude to Y.S. Barash, E.G. Maksimov, L.P. Pitaevskii, A.A. Sobyanin, and G.F. Zharkov for reading the manuscript and remarks.

2.9 Notes

1*. The paper was at first published in *Usp. Fiz. Nauk* **167**, 429 (1997) [*Phys. Usp.* **40**, 407 (1997)].

Some more details concerning the thermoelectric phenomena appeared in *Usp. Fiz. Nauk* **168**, 363 (1998). With account of both these publications, the paper was then included in my book [2] *About Science, Myself, and Others* (Moscow, Fizmatlit, 2003). In the present book the text of the paper is the same, but has some additional small specifications and citations.

2*. In paper [220], the critical field for superheating was calculated to rather a high approximation (for $\varkappa \ll 1$) with the result,

$$\frac{H_{c1}}{H_{cm}} = 2^{-1/4}\varkappa^{-1/2}\left(1 + \frac{15\sqrt{2}}{32}\varkappa + O(\varkappa^2)\right).$$

3*. The talented theoretician physicist A.A. Sobyanin died on 10 June, 1997 at the age of 54. Unfortunately, I am unaware of the fate of his last note mentioned in the footnote on p. 59.

4*. For several years now (beginning from 1995), great attention has been shown in experimental studies of Bose–Einstein condensation (BEC) of rarefied gases at low temperatures. The theoretical analysis has mostly been based on the Gross–Pitaevskii theory (see [221, 234, 237, 240]). The development of this theory, I believe, was significantly influenced by the Ψ-theory of superfluidity. It seems probable that the Ψ-theory of superfluidity [96], both in its original and generalized forms, may also be useful when applied to BEC in gases, particularly in the neighborhood of the λ-point. The thermomechanical circulation effect in superfluids [144, 159, 160] also can be interesting in BEC systems.

5*. In this connection, see the supplement to Chap. 6 in [2].

6*. See also Chap. 19 in [2].

7*. I have seen a statement in the literature that over 50,000 papers have been devoted to HTSC over 10 years.

A number of new interesting and unexpected experimental data concerning high-temperature cuprates have been obtained in recent years [273]. As a result, the mechanism of their superconductivity remains unclear. I hope that its clarification is a matter of the near future although 20 years have already passed since they were discovered (see Chap. 3 in the present book).

8*. According to [231], in superfluid ^{3}He the length $\xi(0) \sim 10^{-5}$cm while, as I pointed out in the text, for ^{4}He the length $\xi(0) \sim 10^{-8}$cm. Clearly, in ^{3}He and in some other cases, the order parameter is not the scalar function Ψ. So, one considers generalizations of the Ψ-theory with Ψ substituted by the corresponding order parameter.

9*. In the preprint [232], some scheme is elaborated that combines the generalized Ψ-theory of superfluidity with BEC theory. I do not see grounds for such a theory but, nevertheless, its analysis seems interesting. In paper [?], the Ψ-theory is somewhat generalized (by taking into account the Van der Waals forces) and compared with an experiment. Unfortunately, the author's conclusions remain obscure to me.

I have to state with regret that the Ψ-theory of superfluidity has not attracted any attention in the new publications known to me devoted to superfluidity in liquid helium, and it is totally ignored in other cases. In fact, superfluidity in liquids is very rarely studied generally now, which can be understood, in particular, in connection with the enthusiasm for BEC in gases. Nevertheless, studies of liquid ^{4}He continue and, for example, I think it would be quite relevant to involve the Ψ-theory of superfluidity for analysis in papers [241,242]. The same relates to studies of superfluidity in ^{3}He, in neutron stars, and in other cases.

10*. I first of all mean the theory in which the parameter Ψ has several components. A good example is here the Ψ-theory developed in application to the superconductor MgB_2 which has two gaps (in this case, two complex scalar functions Ψ_1 and Ψ_2 serve as the order parameter [275]).

11*. This presentation is based on the assumption that the Fermi-liquid model is applicable to cuprates (when they are considered). If, however, the Fermi-liquid notion is inapplicable to cuprates (and possibly to some other superconductors) (see [262] and references therein and [103, 104, 143] in the reference list to Chap. 6 in [2]), a special investigation (both theoretical and experimental) of thermoelectric phenomena in such materials will be necessary. With this fact in mind, it seems to me that the study of thermoelectric effects in superconductors acquires an additional interest.

12*. The problems I see in the field of superconductivity and superfluidity are also discussed in my preprint [257]. They mostly coincide with the topics discussed in the present paper.

13*. It should be borne in mind that the present paper was published in 1997. In presenting it in this book, we have added only a few notes and references to new literature, with the exception of Sect. 2.5 (which is devoted to thermoelectric phenomena). Of course, this does not make the paper as it would have appeared should it be written anew in 2007. However, this concerns only the present state of the physics of superconductivity and not the history of its development, to which (though in the autobiographical aspect) this paper is devoted. So, I hope the small changes to the original text, which somewhat violate its just proportions, prove to be justified.

I take the opportunity to make reference to several new papers (in addition to the already mentioned paper [247]) which have drawn my attention: [248] – superconductivity in thin films with tensions, [249, 250] – vortexes in alternating superconducting and ferromagnetic films, [251, 274, 276] – study of ferromagnetic superconductors, [252] – combinational (Raman) scattering of light in HTSC, [269] – λ-transition analysis in ^{4}He, [260, 261] – superconductivity in MgB_2 and ferromagnetics [274], and superconducting clusters [279]; [263, 278] – discussion of HTSC mechanism, [262] – the elucidation of quasi-particles nature in HTSC material.

14*. In relation to the discussion of thermoelectric effects, I should note that the Ψ-theory [29] was developed under conditions that the electric field $\mathbf{E}$ is either absent or ignored. Thus, the vector potential $\mathbf{A}$ alone was accounted for. When considering thermoelectric effects, in view of necessity of taking into account the field $\mathbf{E}$ there is a need to generalize the Ψ-theory by introducing to it both the vector potential $\mathbf{A}$ and the scalar potential φ.

References

1. V.L. Ginzburg, *The Physics of a Lifetime: Reflections on the Problems and Personalities of 20th Century Physics*, Springer, Berlin, 2001.
2. V.L. Ginzburg, *About Science, Myself and Others*, IOP, Bristol, 2005.
3. *Reminiscences of Landau*, Nauka, Moscow, 1988; *Landau. The Physicist and the Man*, Pergamon Press, Oxford, 1989.
4. L.D. Landau, Zh. Eksp. Teor. Fiz. **11**, 592, 1941; J. Phys. USSR **5**, 71, 1941.
5. P.L. Kapitza, Nature **141**, 74, 1938; Zh. Eksp. Teor. Fiz. **11**, 581, 1941; J. Phys. USSR **4**, 177, 1941; **5**, 59, 1941.
6. G.F. Allen and A.D. Misener, Nature **141**, 75, 1938; Proc. R. Soc. London **172A**, 467, 1939.
7. H. Kamerlingh Onnes, Commun. Phys. Lab. Univ. Leiden **124**c, 1911 (this paper is included as an appendix in a more readily available paper [9]).
8. P.F. Dahl, *Superconductivity. Its Historical Roots and Development from Mercury to the Ceramic Oxides*, American Institute of Physics, New York, 1992. J. de Nobel, Phys. Today **49**(9), 40, 1996.
9. V.L. Ginzburg, *Research on superconductivity (a brief history and outlook for the future)*, Sverkhprovodimost: Fiz., Khim., Tekhnol. 5(1), 1, 1992. [Superconductivity: Phys., Chem., Technol. **5**, 1, 1992].

---

[17] Papers written by the present author or those where he is a co-author are given with titles. This was done naturally with only the purpose of providing additional information, because very little is said about some of these papers in the main text.

10. H. Kamerlingh Onnes, Commun. Phys. Lab. Univ. Leiden **119**a, 1911; Proc. R. Acad. Amsterdam **13**, 1093, 1911.
11. W.H. Keesom, *Helium*, Elsevier, Amsterdam, 1942.
12. F. London, *Superfluids*. Vol. 2. Macroscopic Theory of Superfluid Helium (Wiley & Sons, New York, 1954).
13. R.P. Feynman, *Statistical Mechanics*, (W.A. Benjamin, Reading, MA: 1972).
14. V.L. Ginzburg, *The Present State of the Theory of Superconductivity*. II. *The Microscopic Theory*. Usp. Fiz. Nauk **48**, 26, 1952. [Translation of the main part of the paper, Forschr. Phys. **1**, 101, 1953].
15. R.A. Ogg, Jr. Phys. Rev. **69**, 243, 1946; **70**, 93, 1946.
16. M.R. Schafroth, Phys. Rev. **96**, 1149; 1442, 1954; **100**, 463, 1955; M.R. Schafroth, S.T. Butler and J.M. Blatt, Helv. Phys. Acta **30**, 93, 1957.
17. L.N. Cooper, Phys. Rev. **104**, 1189, 1956.
18. J. Bardeen, L.N. Cooper and J.R. Schrieffer, Phys. Rev. **108**, 1175, 1957.
19. V.L. Ginzburg, *Comments on the Theory of Superconductivity*, Zh. Eksp. Teor. Fiz. **14**, 134, 1944.
20. J. Bardeen, in *Kälterphysik* (Handbuch der Physik **15**), ed. by S. von Flügge, p. 274 (Springer, Berlin, 1956) [Translated into Russian in *Fizika Nizkikh Temperatur (Low-Temperature Physics)*, ed. by A.I. Shal'nikov, p. 679, Inostrannaya Literatura, Moscow, 1959].
21. V.L. Ginzburg, *On Gyromagnetic and Electron-Inertia Experiments with Superconductors*. Zh. Eksp. Teor. Fiz. **14**, 326, 1944.
22. V.L. Ginzburg, *On Thermoelectric Phenomena in Superconductors*, Zh. Eksp. Teor. Fiz. **14**, 177, 1944; J. Phys. USSR **8**, 148, 1944.
23. V.L. Ginzburg, *Light Scattering in Helium* II, Zh. Eksp. Teor. Fiz. **13**, 243, 1943; *A brief report*. J. Phys. USSR **7**, 305, 1943.
24. V.L. Ginzburg, *Superconductivity*. Izd. Akad. Nauk SSSR, Moscow–Leningrad, 1946.
25. W. Meissner and R. Ochsenfeld, Naturwissensch. **21**, 787, 1933.
26. C.J. Gorter and H. Casimir, Physica **1**, 306, 1934; Phys. Z. **35**, 963, 1934.
27. F. London and H. London, Proc. R. Soc. London **149A**, 71, 1935; Physica **2**, 341, 1935.
28. R. Becker, G. Heller and F. Sauter, Z. Phys. **85**, 772, 1933.
29. V.L. Ginzburg and L.D. Landau, *To the Theory of Superconductivity*. Zh. Eksp. Teor. Fiz. **20**, 1064, 1950. [This paper is available in English in L.D. Landau, *Collected Papers*. Pergamon Press, Oxford, 1965].
30. E.M. Lifshitz and L.P. Pitaevskii, *Statisticheskaya Fizika* [Statistical Physics] Pt. 2, *Teoriya Kondensirovannogo Sostoyaniya (Theory of the Condensed State)*, Nauka, Moscow, 1978, 1999 [Translated into English, Pergamon Press, Oxford, 1980].
31. L.P. Gorkov, Zh. Eksp. Teor. Fiz. **36**, 1918, 1959; **37**, 1407, 1959 [Sov. Phys. JETP **9**, 1364, 1959; **10**, 998, 1960].
32. V.L. Ginzburg, *On the Surface Energy and the Behaviour of Small-Sized Superconductors*. Zh. Eksp. Teor. Fiz. **16**, 87, 1946; J. Phys. USSR **9**, 305, 1945.
33. V.L. Ginzburg, *The Present State of the Theory of Superconductivity*. Pt. 1, *Macroscopic theory*, Usp. Fiz. Nauk **42**, 169, 1950.
34. L.D. Landau and E.M. Lifshitz, *Statisticheskaya Fizika (Statistical Physics)* Pt. 1, Fizmatlit, Moscow, 1995, Chap. XIV [Translated into English, Pergamon Press, Oxford, 1980].

35. V.L. Ginzburg, *The Theory of Ferroelectric Phenomena*. Usp. Fiz. Nauk **38**, 400, 1949.

36. V.L. Ginzburg, *On the Theory of Superconductivity*. Nuovo Cimento **2**, 1234, 1955.

37. P.-G. de Gennes, *Superconductivity of Metals and Alloys*, W.A. Benjamin, New York, 1966 [Translated into Russian, Mir, Moscow, 1968].

38. D. Saint-James, G. Sarma and E.J. Thomas, *Type II Superconductivity*, Pergamon Press, Oxford, 1969 [Translated into Russian, Mir, Moscow, 1970].

39. D.R. Tilley and J. Tilley, *Superfluidity and Superconductivity*, 2nd edn., Adam Hilger, Bristol, 1986.

40. V.V. Schmidt, *Vvedenie v Fiziku Sverkhprovodnikov (Introduction to the Physics of Superconductors)*, 2nd edn., MTsNMO, Moscow, 1997 [Translated into English, *The Physics of Superconductors: Introduction to Fundamental and Applications*, Springer, Berlin, 1997].

41. A.A. Abrikosov, *Osnovy Teorii Metallov (Fundamental Principles of the Theory of Metals)*, Fizmatlit, Moscow, 1987 [Translated into English, Elsevier, New York, 1988].

42. E. Weinan, Phys. Rev. B **50**, 1126, 1994; Physica **77D**, 383, 1994.

43. H. Sakaguchi, Prog. Theor. Phys. **93**, 491, 1995.

44. M.V. Bazhenov, M.I. Rabinovich and A.L. Fabrikant, Phys. Lett. **163A**, 87, 1992; R.J. Deissler and H.R. Brand, Phys. Rev. Lett. **72**, 478,1994; J.M. Soto-Crespo, N.N. Akhmediev and V.V. Afanasjev, Optics Commun. **118**, 587, 1995.

45. F. Bethuel, H. Brezis and F. Helein, *Ginzburg–Landau Vortices*, Birkhauser, Boston, 1994.

46. D.A. Kirzhnits, Usp. Fiz. Nauk **125**, 169, 1978 [Sov. Phys. Usp. **21**, 470, 1978].

47. V.L. Ginzburg, (*Bibliography of Soviet Scientists*) (Physics Series **21**), Nauka, Moscow, 1978.

48. V.L. Ginzburg, *On the non-linearity of electromagnetic processes in superconductors*, J. Phys. USSR **11**, 93, 1947.

49. V.L. Ginzburg, *The theory of superfluidity and critical velocity in helium II*, Dokl. Akad. Nauk SSSR **69**, 161, 1949.

50. L.D. Landau, Zh. Eksp. Teor. Fiz. **7**, 19, 627, 1937; Phys. Z. Sowjetunion **11**, 26; 545, 1937.

51. C.N. Yang, Rev. Mod. Phys. **34**, 694, 1962.

52. O. Penrose and L. Onsager, Phys. Rev. **104**, 576, 1956.

53. J.W. Lynn (ed.), *High Temperature Superconductivity*, Springer, Berlin, 1990.

54. V.L. Ginzburg, *Theories of superconductivity (a few remarks)*, Helv. Phys. Acta **65**, 173, 1992.

55. V.L. Ginzburg, *To the macroscopic theory of superconductivity*, Zh. Eksp. Teor. Fiz. **29**, 748, 1955 [Sov. Phys. JETP **2**, 589, 1956].

56. C.J. Boulter and J.O. Indeken, Phys. Rev. B **54**, 12407, 1996; J.M. Mishonov, J. Physique **51**, 447, 1990.

57. V.L. Ginzburg, *An experimental manifestation of instability of the normal phase in superconductors*, Zh. Eksp. Teor. Fiz. **31**, 541, 1956 [Sov. Phys. JETP **4**, 594, 1957].

58. D. Shoenberg, *Superconductivity*, 3rd edn., Cambridge University Press, Cambridge, 1965 [Russian translation of the previous edition, IL, Moscow, 1955].

59. A.A. Abrikosov, Zh. Eksp. Teor. Fiz. **32**, 1442, 1957 [Sov. Phys. JETP **5**, 1174, 1957]; Dokl. Akad. Nauk SSSR **86**, 489, 1952.

60. V.L. Ginzburg, *On the behaviour of superconducting films in a magnetic field*, Dokl. Akad. Nauk SSSR **83**, 385, 1952.

61. V.P. Silin, Zh. Eksp. Teor. Fiz. **21**, 1330, 1951.

62. V.L. Ginzburg, *On the destruction and the onset of superconductivity in a magnetic field*, Zh. Eksp. Teor. Fiz. **34**, 113, 1958 [Sov. Phys. JETP **7**, 78, 1958].

63. V.L. Ginzburg, *Critical current for superconducting films*, Dokl. Akad. Nauk SSSR **118**, 464, 1958.

64. V.L. Ginzburg, *On the behaviour of superconductors in a high-frequency field*, Zh. Eksp. Teor. Fiz. **21**, 979, 1951.

65. J. Bardeen, Phys. Rev. **94**, 554, 1954.

66. V.L. Ginzburg, *Some remarks concerning the macroscopic theory of superconductivity*, Zh. Eksp. Teor. Fiz. **30**, 593, 1956 [Sov. Phys. JETP **3**, 621, 1956].

67. V.L. Ginzburg, *To the macroscopic theory of superconductivity valid at all temperatures*, Dokl. Akad. Nauk SSSR **110**, 358, 1956.

68. V.L. Ginzburg, *On comparison of the macroscopic theory of superconductivity with experimental data*, Zh. Eksp. Teor. Fiz. **36**, 1930, 1959 [Sov. Phys. JETP **36**, 1372, 1959].

69. V.L. Ginzburg, *Allowance for the effect of pressure in the theory of second-order phase transitions (as applied to the case of superconductivity)*, Zh. Eksp. Teor. Fiz. **44**, 2104, 1963 [Sov. Phys. JETP **17**, 1415, 1963].

70. F. London, *Superfluids*. Vol. 1, *Macroscopic Theory of Superconductivity*, Wiley, New York, 1950.

71. W. Buckel, *Supraleitung*, Physik, Weinheim, 1972 [Translated into English, *Superconductivity: Fundamentals and Applications*, VCH, Weinheim, 1991; translation into Russian, Mir, Moscow, 1975].

72. V.L. Ginzburg, *Magnetic flux quantization in a superconducting cylinder*, Zh. Eksp. Teor. Fiz. **42**, 299, 1962 [Sov. Phys. JETP **15**, 207, 1962].

73. J. Bardeen, Phys. Rev. Lett. **7**, 162, 1961.

74. J.B. Keller and B. Zumino, Phys. Rev. Lett. **7**, 164, 1961.

75. V.L. Ginzburg, *On account of anisotropy in the theory of superconductivity*, Zh. Eksp. Teor. Fiz. **23**, 236, 1952.

76. V.L. Ginzburg, *Ferromagnetic superconductors*, Zh. Eksp. Teor. Fiz. **31**, 202, 1956 [Sov. Phys. JETP **4**, 153, 1957].

77. L.N. Bulaevskii, in *Superconductivity, Superdiamagnetism, Superfluidity*, ed. by V.L. Ginzburg, p. 69, Mir, Moscow, 1987.

78. M. Laue, Ann. Phys. (Leipzig) **3**, 31, 1948.

79. M. Laue, *Theorie der Supraleitung*, Springer, Berlin, 1949.

80. V.L. Ginzburg, *Some questions of the theory of electric fluctuations*, Usp. Fiz. Nauk **46**, 348, 1952.

81. V.V. Schmidt, Pis'ma Zh. Eksp. Teor. Fiz. **3**, 141, 1966 [JETP Lett. **3**, 89, 1966].

82. H. Schmidt, Z. Phys. **216**, 336, 1968.

83. A. Schmid, Phys. Rev. **180**, 527, 1969.

84. L.G. Aslamazov and A.I. Larkin, Fiz. Tverd. Tela (Leningrad) **10**, 1104, 1968 [Sov. Phys. Solid State **10**, 875, 1968].

85. M. Tinkham, *Introduction to Superconductivity*, 2nd edn., McGraw-Hill, New York, 1996.

86. V.L. Ginzburg, *Several remarks on second-order phase transitions and micro-scopic theory of ferroelectrics*, Fiz. Tverd. Tela (Leningrad) **2**, 2031, 1960 [Sov. Phys. Solid State **2**, 1824, 1961].

87. A.P. Levanyuk, Zh. Eksp. Teor. Fiz. **36**, 810, 1959 [Sov. Phys. JETP **9**, 571, 1959].

88. A.Z. Patashinskii and V.L. Pokrovskii, *Fluktuatsionnaya Teoriya Fazovykh Perekhodov*, 2nd edn. [*Fluctuation Theory of Phase Transitions*], Nauka, Moscow, 1982 [Translation into English, Pergamon Press, Oxford, 1979].

89. V.L. Ginzburg, A.P. Levanyuk and A.A. Sobyanin, *Comments on the region of applicability of the Landau theory for structural phase transitions*, Ferro-electrics **73**, 171, 1983.

90. L.N. Bulaevskii, V.L. Ginzburg and A.A. Sobyanin, *Macroscopic theory of superconductors with small coherence length*, Zh. Eksp. Teor. Fiz. **94**, 356, 1988 [Sov. Phys. JETP **68**, 1499, 1989]; Usp. Fiz. Nauk **157**, 539, 1989 [Sov. Phys. Usp. **32**, 1277, 1989]; Physica C **152**, 378, 1988; **153–155**, 1617, 1988.

91. A.A. Sobyanin and A.A. Stratonnikov, Physica C **153–155**, 1680, 1988.

92. L.P. Gorkov and N.B. Kopnin, Usp. Fiz. Nauk **156**, 117, 1988 [Sov. Phys. Usp. **31**, 850, 1988].

93. A.S. Alexandrov and N.F. Mott, *High Temperature Superconductors and Other Superfluids*, Taylor & Francis, London, 1994.

94. V.L. Ginzburg, *On the Ψ-theory of high temperature superconductivity*. In: Proc. 18th Int. Conf. on Low Temperature Physics, 20–26 August, 1987, Kyoto, Japan, LT-18; Japan J. Appl. Phys. **26** (Suppl. 26-3), 2046, 1987.

95. E.A. Andryushin, V.L. Ginzburg and A.P. Silin, *On boundary conditions in the macroscopic theory of superconductivity*, Usp. Fiz. Nauk **163**, 105, 1993 [Phys. Usp. **36**, 854, 1993].

96. V.L. Ginzburg and L.P. Pitaevskii, *On the theory of superfluidity*, Zh. Eksp. Teor. Fiz. **34**, 1240, 1958 [Sov. Phys. JETP **7**, 858, 1958].

97. M. Cyrot, Reps. Prog. Phys. **36**, 103, 1973.

98. L.P. Pitaevskii, Zh. Eksp. Teor. Fiz. **37**, 1794, 1959 [Sov. Phys. JETP **37**, 1267, 1960].

99. G.E. Volovik and L.P. Gorkov Zh. Eksp. Teor. Fiz. **88**, 1412, 1985 [Sov. Phys. JETP **61**, 843, 1985].

100. J.F. Annett, Adv. Phys. **39**, 83, 1990; Contemp. Phys. **36**, 423, 1995.

101. M. Sigrist and K. Ueda, Rev. Mod. Phys. **63**, 239, 1991.

102. J.A. Sauls, Adv. Phys. **43**, 113, 1994; V.M. Edelstein, J. Phys. Cond. Mat. **8**, 339, 1996.

103. D.L. Cox and M.B. Maple, Phys. Today **48**(2), 32, 1995.

104. Yu.S. Barash, A.V. Galaktionov and A.D. Zaikin, Phys. Rev. **52B**, 665, 1995; Yu.S. Barash and A.A. Svidzinsky, Phys. Rev. **53B**, 15254, 1996; Phys. Rev. Lett. **77**, 4070, 1996.

105. Yu.S. Barash and V.L. Ginzburg, Usp. Fiz. Nauk **116**, 5, 1975; **143**, 346, 1984 [Sov. Phys. Usp. **18**, 305, 1976; **27**, 467, 1984].

106. V.L. Ginzburg, *The surface energy associated with a tangential velocity dis-continuity in helium II*, Zh. Eksp. Teor. Fiz. **29**, 254, 1955 [Sov. Phys. JETP **2**, 170, 1956].

107. G.A. Gamtsemlidze, Zh. Eksp. Teor. Fiz. **34**, 1434, 1958 [Sov. Phys. JETP **34**, 992, 1958].

108. V.L. Ginzburg and A.A. Sobyanin, *Superfluidity of helium II near the λ-point*, Usp. Fiz. Nauk **120**, 153, 1976 [Sov. Phys. Usp. **19**, 773, 1976].

109. V.L. Ginzburg and A.A. Sobyanin, *On the theory of superfluidity of helium II near the λ-point*, J. Low Temp. Phys. **49**, 507, 1982.

110. V.L. Ginzburg and A.A. Sobyanin, *Superfluidity of helium II near the λ-point*. In *Superconductivity, Superdiamagnetism, Superfluidity*, ed. by V.L. Ginzburg, p. 242, Mir, Moscow, 1987.

111. A.A. Sobyanin, Zh. Eksp. Teor. Fiz. **63**, 1780, 1972 [Sov. Phys. JETP **36**, 941, 1972].

112. V.L. Ginzburg and A.A. Sobyanin, *Superfluidity of helium II near the λ-point*, Usp. Fiz. Nauk **154**, 545, 1988 [Sov. Phys. Usp. **31**, 289, 1988]; Japan J. Appl. Phys. **26** (Suppl. 26-3), 1785, 1987.

113. L.S. Goldner, N. Mulders and G. Ahlers, J. Low Temp. Phys. **89**, 131, 1992.

114. V.L. Ginzburg and A.A. Sobyanin, *Can liquid molecular hydrogen be superfluid?* Pis'ma Zh. Eksp. Teor. Fiz. **15**, 343, 1972 [JETP Lett. **15**, 242, 1972].

115. G.P. Collins, Phys. Today **49**(8), 18, 1996.

116. R.P. Feynman, in *Progress in Low Temperature Physics*, Vol. 1, ed. by C.J. Gorter, p. 1, North-Holland, Amsterdam, 1955.

117. Yu.G. Mamaladze, Zh. Eksp. Teor. Fiz. **52**, 729, 1967 [Sov. Phys. JETP **25**, 479, 1967]; Phys. Lett. **A27**, 322, 1968.

118. V. Dohm and R. Haussmann, Physica B **197**, 215, 1994.

119. V.L. Ginzburg and A.A. Sobyanin, *Structure of vortex filament in helium II near the λ-point*, Zh. Eksp. Teor. Fiz. **82**, 769, 1982 [Sov. Phys. JETP **55**, 455, 1982].

120. L.P. Pitaevskii, Zh. Eksp. Teor. Fiz. **35**, 408, 1958 [Sov. Phys. JETP **35**, 282, 1959].

121. W.H. Zurek, Nature **382**, 296, 1996.

122. F.M. Gasparini and I. Rhee, Prog. Low Temp. Phys. **13**, 1, 1992.

123. L.V. Mikheev and M.E. Fisher, J. Low Temp. Phys. **90**, 119, 1993.

124. W. Zimmermann, Contemp. Phys. **37**, 219, 1996.

125. M. Chan, N. Mulders and J. Reppy, Phys. Today **49**(8), 30, 1996.

126. E.F. Burton, H.G. Smith and J.O. Wilhelm, *Phenomena at the Temperature of Liquid Helium*, Reinhold, New York, 1940.

127. R.J. Donnelly, Phys. Today **48**(7), 30, 1995.

128. V.L. Ginzburg and G.F. Zharkov, *Thermoelectric effects in superconductors*, Usp. Fiz. Nauk **125**, 19, 1978 [Sov. Phys. Usp. **21**, 381, 1978].

129. V.L. Ginzburg, *Thermoelectric effects in the superconducting state*, Usp. Fiz. Nauk **161**(2), 1, 1991 [Sov. Phys. Usp. **34**, 101, 1991].

130. B.T. Geilikman and V.Z. Kresin, *Kineticheskie i Nestatsionarnye Yavleniya v Sverkhprovodnikakh [Kinetic and Nonsteady-State Phenomena in Superconductors]*, Nauka, Moscow, 1972 [Translated into English, J. Wiley, New York, 1974]; Zh. Eksp. Teor. Fiz. **34**, 1042, 1958 [Sov. Phys. JETP **7**, 721, 1958].

131. A.I. Anselm, *Introduction to the Theory of Superconductors*, Chap. 8, Gos. Izd. Fiz. Mat. Lit., Moscow–Leningrad, 1962; K. Seeger, *Semiconductor Physics*, Springer, Heidelberg, 1997.

132. V.L. Ginzburg, *Convective heat transfer and other thermoelectric effects in high-temperature superconductors*, Pis'ma Zh. Eksp. Teor. Fiz. **49**, 50, 1989 [JETP Lett. **49**, 58, 1989].

133. L.Z. Kon, Zh. Eksp. Teor. Fiz. **70**, 286, 1976 [Sov. Phys. JETP **43**, 149, 1976]; Fiz. Tverd. Tela (Leningrad), **19**, 3695, 1977 [Sov. Phys. Solid State **19**, 2160, 1977]; D.F. Digor, L.Z. Kon and V.A. Moskalenko, Sverkhprovodimost': Fiz.,

Khim., Tekhnol. **3**, 2485, 1990 [Superconductivity: Phys., Chem., Technol. **3**, 1703, 1990].

134. B. Arfi et al., Phys. Rev. Lett. **60**, 2206, 1988; Phys. Rev. **39B**, 8959, 1989; P.J. Hirschfeld, Phys. Rev. **B37**, 9331, 1988.

135. A. Jezowski et al., Helv. Phys. Acta **61**, 438, 1988; Phys. Lett. **138A**, 265, 1989.

136. J.L. Cohn et al., Phys. Rev. **B45**, 13144, 1992; R.C. Yu et al., Phys. Rev. Lett. **69**, 1431, 1992.

137. J.L. Cohn et al., Phys. Rev. Lett. **71**, 1657, 1993.

138. P.J. Hirschfeld and W.O. Putikka, Phys. Rev. Lett. **77**, 3909, 1996.

139. V.L. Ginzburg, *Thermoelectric effects in Superconductors*, J. Supercond. **2**, 323, 1989; Physica C **162–164**, 277, 1989; Supercond. Sci. Technol. **4**, 1, 1991.

140. V.L. Ginzburg and G.F. Zharkov, *Thermoelectric effect in anisotropic superconductors*, Pis'ma Zh. Eksp. Teor. Fiz. **20**, 658, 1974 [JETP Lett. **20**, 302, 1974].

141. P.M. Selzer and W.M. Fairbank, Phys. Lett. **A48**, 279, 1974.

142. Yu.M. Gal'perin, V.L. Gurevich and V.N. Kozub, Zh. Eksp. Teor. Fiz. **66**, 1387, 1974 [Sov. Phys. JETP **39**, 680, 1974].

143. J.C. Garland and D.J. van Harlingen, Phys. Lett. **A47**, 423, 1974.

144. V.L. Ginzburg, G.F. Zharkov and A.A. Sobyanin, *Thermoelectric phenomena in superconductors and thermomechanical circulation effect in a superfluid liquid*, Pis'ma Zh. Eksp. Teor. Fiz. **20**, 223, 1974 [JETP Lett. **20**, 97, 1974]; V.L. Ginzburg, A.A. Sobyanin and G.F. Zharkov, Phys. Lett. **A87**, 107, 1981.

145. D.J. van Harlingen, Physica, **B109–110**, 1710, 1982.

146. A.M. Gerasimov et al., Czechoslovak J. Phys. **46**, S.2, 633 (LT21), 1996; Sverkhprov.: Fiz., Khim., Tekhnol. **8**, 634, 1995; J. Low Temp. Phys. **106**, 591, 1997.

147. R.M. Arutyunyan and G.F. Zharkov, Zh. Eksp. Teor. Fiz. **83**, 1115, 1982 [Sov. Phys. JETP **56**, 632, 1982]; J. Low. Temp. Phys. **52**, 409, 1983; Phys. Lett. **A96**, 480, 1983.

148. V.L. Ginzburg, G.F. Zharkov and A.A. Sobyanin, *Thermoelectric current in a superconducting circuit*, J. Low Temp. Phys. **47**, 427, 1982; **56**, 195, 1984.

149. V.L. Ginzburg and G.F. Zharkov, *Thermoelectric effect in hollow superconducting cylinders*, J. Low Temp. Phys. **92**, 25, 1993.

150. V.L. Ginzburg and G.F. Zharkov, *Thermoelectric effects in superconducting state*, Physica **235C–240C**, 3129, 1994.

151. R.M. Arutyunyan, V.L. Ginzburg and G.F. Zharkov, *Vortices and thermoelectric effect in a hollow superconducting cylinder*, Zh. Eksp. Teor. Fiz. **111**, 2175, 1997 [JETP **84**, 1186, 1997]; Usp. Fiz. Nauk **167**, 457, 1997; [Phys. Usp. **40**, 435, 1997].

152. B.A. Mattoo and Y. Singh, Prog. Theor. Phys. **70**, 51, 1983.

153. R.P. Huebener, A.V. Ustinov and V.K. Kaplunenko, Phys. Rev. **B42**, 4831, 1990.

154. A.V. Ustinov, M. Hartmann and R.P. Huebener, Europhys. Lett. **13**, 175, 1990.

155. W.F. Vinen, J. Phys. C: Solid State Phys. **4**, 1287, 1971; **8**, 101, 1975.

156. I.L. Fabelinskii, Usp. Fiz. Nauk **164**, 897, 1994 [Phys. Usp. **37**, 821, 1994].

157. V.L. Ginzburg and A.A. Sobyanin, *Use of second sound to investigate the inhomogeneous density distribution of the superfluid part of helium II near the λ-point*, Pis'ma Zh. Eksp. Teor. Fiz. **17**, 698, 1973 [JETP Lett. **17**, 483, 1973].

158. V.L. Ginzburg, *On the superfluid flow induced by crossed electric and magnetic fields*, Fiz. Nizk. Temp. **5**, 299, 1979 [Sov. J. Low Temp. Phys. **5**, 1979].

159. V.L. Ginzburg and A.A. Sobyanin, *Circulation effect and quantum interference phenomena in a nonuniformly heated toroidal vessel with superfluid helium*, Zh. Eksp. Teor. Fiz. **85**, 1606, 1983 [Sov. Phys. JETP **58**, 934, 1983].

160. G.A. Gamtsemlidze and M.I. Mirzoeva, Zh. Eksp. Teor. Fiz. **79**, 921, 1980; **84**, 1725, 1983 [Sov. Phys. JETP **52**, 468, 1980; **57**, 1006, 1983].

161. V.L. Ginzburg, *Second sound, the convective heat transfer mechanism, and exciton excitations in superconductors*, Zh. Eksp. Teor. Fiz. **41**, 828, 1961 [Sov. Phys. JETP **14**, 594, 1962].

162. V.L. Ginzburg, A.A. Gorbatsevich, Yu.V. Kopaev and B.A. Volkov, *On the problem of superdiamagnetism*, Solid State Commun. **50**, 339, 1984.

163. V.L. Ginzburg, *Theory of superdiamagnetics*, Pis'ma Zh. Eksp. Teor. Fiz. **30**, 345, 1979 [JETP Lett. **30**, 319, 1979].

164. A.A. Gorbatsevich, Zh. Eksp. Teor. Fiz. **95**, 1467, 1989 [Sov. Phys. JETP **68**, 847, 1989].

165. V.L. Ginzburg and D.A. Kirzhnits, *On the superfluidity of neutron stars*, Zh. Eksp. Teor. Fiz. **47**, 2006, 1964 [Sov. Phys. JETP **20**, 1346, 1965].

166. V.L. Ginzburg and D.A. Kirzhnits, *Superconductivity in white dwarfs and pulsars*, Nature **220**, 148, 1968.

167. V.L. Ginzburg and G.F. Zharkov, *Superfluidity of the cosmological neutrino sea*, Pis'ma Zh. Eksp. Teor. Fiz. **5**, 275, 1967 [JETP Lett. **5**, 223, 1967].

168. V.L. Ginzburg, *Superfluidity and superconductivity in the Universe*, Usp. Fiz. Nauk **97**, 601, 1969 [Sov. Phys. Usp. **12**, 241, 1970]; Physica **55**, 207, 1971.

169. I.E. Tamm, Phys. Z. Sowjetunion **1**, 733, 1932.

170. V.L. Ginzburg and D.A. Kirzhnits, *On the superconductivity of electrons at the surface levels*, Zh. Eksp. Teor. Fiz. **46**, 397, 1964 [Sov. Phys. JETP **19**, 269, 1964].

171. L.N. Bulaevskii and V.L. Ginzburg, *On the possibility of the existence of surface ferromagnetism*, Fiz. Met. Metalloved. **17**, 631, 1964 [Phys. Metals Metallography **17**, 1964].

172. P.C. Hohenberg, Phys. Rev. **158**, 383, 1967.

173. V.L. Ginzburg and D.A. Kirzhnits, *The question of high-temperature and surface superconductivity*, Dokl. Akad. Nauk SSSR **176**, 553, 1967 [Sov. Phys. Dokl. **12**, 880, 1968].

174. V.L. Ginzburg, *On two-dimensional superconductors*, Phys. Scripta **27**, 76, 1989.

175. V.L. Ginzburg and D.A. Kirzhnits (Eds.), *Problema Vysokotemperaturnoi Sverkhprovodimosti* [*High-Temperature Superconductivity*], Nauka, Moscow, 1977 [English translation, Consultants Bureau, New York, 1982].

176. V.L. Ginzburg, *On the electrodynamics of two-dimensional (surface) superconductors*, Essays in Theoretical Physics (in Honour of Dirk ter Haar), ed. by W.E. Parry, p. 43, Oxford, Pergamon Press, 1984.

177. L.N. Bulaevskii, V.L. Ginzburg and G.F. Zharkov, *Behavior of surface (two-dimensional) superconductors and of a very thin superconducting film in a magnetic field*, Zh. Eksp. Teor. Fiz. **85**, 1707, 1983 [Sov. Phys. JETP **58**, 994, 1983].

178. R.F. Frindt, Phys. Rev. Lett. **28**, 299, 1972.

179. V.L. Ginzburg, "Bill Little and high temperature superconductivity." In *From High-Temperature Superconductivity to Microminiature Refrigeration*, ed. by B. Cabrera, H. Gutfreund and V. Kresin, Plenum, New York, 1996.

180. J.G. Bednorz and K.A. Müller, Z. Phys. **B64**, 189, 1986.

181. M.K. Wu et al., Phys. Rev. Lett. **58**, 908, 1987.

182. Proc. Int. Conf. on Materials and Mechanisms of Superconductivity, Kanazawa (Japan), July, *High Temperature Superconductors III (Conf. M^2HTSC III)*; Physica C **185**, 1991.

183. W.A. Little, Phys. Rev. **A134**, 1416, 1964; Sci. Am. **212**(2), 21, 1965.

184. V.L. Ginzburg, *Concerning surface superconductivity*, Zh. Eksp. Teor. Fiz. **47**, 2318, 1964 [Sov. Phys. JETP **20**, 1549, 1964]; Phys. Lett. **13**, 101, 1964.

185. V.L. Ginzburg, *The problem of high-temperature superconductivity*, Usp. Fiz. Nauk **95**, 91, 1968 [Contemp. Phys. **9**, 355, 1968]; **101**, 185, 1970 [Sov. Phys. Usp. **13**, 335, 1971].

186. V.L. Ginzburg, *The problem of high-temperature superconductivity*, Contemp. Phys. **9**, 355, 1968.

187. V.L. Ginzburg, *Manifestation of the exciton mechanism in the case of granulated superconductors*, Pis'ma Zh. Eksp. Teor. Fiz. **14**, 572, 1971 [JETP Lett. **14**, 396, 1971].

188. V.L. Ginzburg, *The problem of high-temperature superconductivity*, Annu. Rev. Mat. Sci. **2**, 663, 1972.

189. V.L. Ginzburg, *High temperature superconductivity*, J. Polymer Sci. C **29**(3), 133, 1970.

190. M.L. Cohen and P.W. Anderson, *Superconductivity in d- and f-Band Metals*, (American Institute of Physics Conf. Ser. 4), ed. by D.H. Duglass, p. 17, New York, American Institute of Physics, 1972.

191. D.A. Kirzhnits, Usp. Fiz. Nauk **119**, 357, 1976 [Sov. Phys. Usp. **19**, 530, 1976].

192. V.L. Ginzburg, *Once again about high-temperature superconductivity*, Contemp. Phys. **33**, 15, 1992.

193. O.V. Dolgov, D.A. Kirzhnits and E.G. Maksimov, Rev. Mod. Phys. **53**, 81, 1981.

194. O.V. Dolgov and E.G. Maksimov, Usp. Fiz. Nauk **138**, 95, 1982 [Sov. Phys. Usp. **25**, 688, 1982].

195. G.M. Eliashberg, Zh. Eksp. Teor. Fiz. **38**, 966; **39**, 1437, 1960 [Sov. Phys. JETP **11**, 696, 1960; **12**, 1000, 1961].

196. V.L. Ginzburg, *High-temperature superconductivity: some remarks*, Prog. Low Temp. Phys. **12**, 1, 1989.

197. V.L. Ginzburg, *High-temperature superconductivity: its possible mechanisms*, Physica **C209**, 1, 1993.

198. V.L. Ginzburg, *High-temperature superconductivity*, Energiya (9), 2, 1984.

199. V.L. Ginzburg, *John Bardeen and the theory of superconductivity*, 2001 (see [1], p. 451); J. Supercond. **4**, 327, 1986.

200. D.M. Ginsberg (Ed.), *Physical Properties of High-Temperature Superconductors* I, Vol. 1, World Scientific, Singapore, 1989 [Several other volumes of this series appeared later].

201. V.L. Ginzburg and E.G. Maksimov, *Mechanisms and models of high temperature superconductors*, Physica **235C–240C**, 193, 1994.

202. D. Shimada et al., Phys. Rev. **B51**, 16495, 1995.

203. E.G. Maksimov, J. Supercond. **8**, 433, 1995.

204. E.G. Maksimov, S.U. Savrasov, D.U. Savrasov and O.V. Dolgov, Solid State Commun. **106**(7), 409, 1998.
205. V.L. Ginzburg and E.G. Maksimov, *On possible mechanisms of high-tempera-ture superconductivity (review)*, Sverkhprovodimost': Fiz., Khim., Tekhnol. **5**, 1543, 1992 [Superconductivity: Phys. Chem. Technol. **5**, 1505, 1992].
206. E. Abrahams et al., Phys. Rev. **B52**, 1271, 1995.
207. R. Fehrenbacher and M.R. Norman, Phys. Rev. Lett. **74**, 3884, 1995.
208. C. O'Donovan and J.R. Carbotte, Physica **252C**, 87, 1995.
209. I. Bozovic et al., J. Supercond. **7**, 187, 1994.
210. M. Cyrot and D. Pavina, *Introduction to Superconductivity and High-T_c Ma-terials*, World Scientific, Singapore, 1992.
211. V.L. Ginzburg and E.A. Andryushin, *Sverkhprovodimost' [Superconductivity]*, 2nd edn., revised and amended, Al'fa-M, Moscow, 2006 [English translation, World Scientific, Singapore, 2005].
212. G.R. Lubkin, Phys. Today **49**(3), 48, 1996.
213. V.L. Ginzburg, *On heat transfer (thermal conductivity) and thermoelectric effect in superconducting state*, Usp. Fiz. Nauk **168**, 363, 1998 [Phys. Usp. **41**, 307, 1998].
214. N.M. Plakida, *Vysokotemperaturnye Sverkhprovodniki [High-Temperature Su-perconductors]*, Mezhdunarodnaya Programma Obrazovaniya, Moscow, 1996 [Translated into English, *High-Temperature Superconductivity: Experiment and Theory*, Springer, Berlin, 1995].
215. A.A. Golubov et al., Physica **235C–240C**, 2383, 1994.
216. R. Combescot and X. Leyronas, Phys. Rev. Lett. **75**, 3732, 1995.
217. M.B. Maple, Physica **215B**, 110, 1995; Physica **215B**, 127, 1995.
218. I. Bozovic and J.N. Eckstein, in *Physical Properties of High-Temperature Su-perconductors*, Vol. 5, ed. by D.M. Ginsberg, World Scientific, Singapore, 1996; see also [118] in Chap. 6 in [2].
219. P.J. Ford and G.A. Saunders, Contemp. Phys. **38**, 63, 1997.
220. A.J. Dolgert et al., Phys. Rev. **B53**, 5650, 1996; **56**, 2883, 1997.
221. F. Dalforo et al., Rev. Mod. Phys. **71**, 463, 1999.
222. P.G. Klemens, Proc. Phys. Soc. London **A66**, 576, 1953; P.G. Klemens, in *Handbuch der Physik* **14**, ed. by S. Flügge, p. 198, Springer, Berlin, 1956 [Rus-sian translation in *Fizika Nizkikh Temperatur [Low-Temperature Physics]*, ed. by A.I. Shal'nikov, IL, Moscow, 1959].
223. N.K. Fedorov, Solid State Commun. **106**, 177, 1998.
224. L.D. Landau and E.M. Lifshitz, *Electrodinamika Sploshnykh Sred* [Electro-dynamics of Continuous Media] 27, Nauka, Moscow, 1992 [Translated into English, Pergamon Press, Oxford, 1984]
225. K. Krishana, J.M. Harris and N.P. Ong, Phys. Rev. Lett. **75**, 3529, 1995.
226. H.J. Mamin, J. Clarke and D.J. van Harlingen, Phys. Rev. Lett. **51**, 1480, 1983.
227. A.M. Gulian and G.F. Zharkov, in *Thermoelectricity in Metallic Conductors*, ed. by F.J. Blatt and P.A. Schroeder, Plenum, New York, 1978 [The Russian text, *Kratk. Soobshch. Fiz.* FIAN (11), 21, 1977].
228. A.M. Gulian and G.F. Zharkov, *Nonequilibrium Electrons and Phonons in Superconductors*, Kluwer, Plenum, New York, 1999.
229. G.F. Zharkov, Zh. Eksp. Teor. Fiz. **122**, 600, 2002 [JETP **95**, 517, 2002]; J. Low Temp. Phys. **130**, 45, 2003; Usp. Fiz. Nauk **174**, 1012, 2004 [Phys. Usp. **47**, 944, 2004].

230. G.F. Zharkov, Zh. Eksp. Teor. Fiz. **34**, 412, 1958; **37**, 1784, 1959 [Sov. Phys. JETP **7**, 278, 1958; **10**, 1257, 1959].
231. R.W. Simmonds et al., Phys. Rev. Lett. **87**, 035301, 2001.
232. T. Fliessbach, Effective Ginzburg–Landau model for superfluid ^{4}He, Preprint cond-mat/0106237, 2001.
233. M.V. Sadovskii, Usp. Fiz. Nauk **171**, 539, 2001 [Phys. Usp. **44**, 515, 2001]; V.M. Loktev et al., Phys. Rep. **349**, 1, 2001.
234. E. Timmermans, Contemp. Phys. **42**, 1, 2001.
235. V.L. Ginzburg, Unthought, undone... Preprint 34, FIAN, Moscow, 2001.
236. S. Mo, J. Hove and A. Sudbo, Phys. Rev. **B65**, 104501, 2002.
237. N.B. Kopnin, J. Low Temp. Phys. **129**, 219, 2002.
238. V.G. Kogan and V.L. Pokrovsky, Phys. Rev. Lett. **90**, 067004, 2003.
239. H. Kleinert and A.M. Schakel, Phys. Rev. Lett. **90**, 097001, 2003.
240. L. Pitaevskii and S. Stringari, *Bose–Einstein Condensation*, Clarendon, Oxford, 2003.
241. D. Murphy et al., Phys. Rev. Lett. **90**, 025301, 2003.
242. T. Chui et al., Phys. Rev. Lett. **90**, 085301, 2003.
243. P. Lipavsky et al., Phys. Rev. **B65**, 012507, 2001.
244. I. Ussishkin, S.L. Sondhi and D.A. Huse, Phys. Rev. Lett. **89**, 287001, 2002.
245. E. Boaknin et al., Phys. Rev. Lett. **90**, 117003, 2003.
246. Y.M. Galperin et al., Phys. Rev. **B65**, 064531, 2002.
247. I. Bozovic et al., Nature **422**, 873, 2003.
248. I. Bozovic et al., Phys. Rev. Lett. **89**, 107001, 2002.
249. Y. Chen et al., Phys. Rev. Lett. **89**, 217001, 2002.
250. E.W. Carlson, A.H.C. Neto and D.K. Campbell, Phys. Rev. Lett. **90**, 087001, 2003.
251. J.Y. Gu et al., Phys. Rev. Lett. **90**, 087001, 2003.
252. O.V. Misochko, *Usp. Fiz. Nauk* **173**, 385, 2003 [Phys. Usp. **46**, 373, 2003].
253. C.P. Chu et al., J. Supercond. **13**, 679, 2000.
254. A.B. Shicll et al., J. Supercond. **13**, 687, 2000.
255. Schmideshoff et al., J. Supercond. **13**, 847, 2000.
256. E.G. Maksimov, Usp. Fiz. Nauk **170**, 1033, 2000 [Phys. Usp. **43**, 965, 2000].
257. K.P. Mooney and F.M. Gasparini, J. Low Temp. Phys. **126**, 247, 2002.
258. M.B. Walker and R.V. Samokhin, Phys. Rev. Lett. **88**, 207001, 2002.
259. M. Cuoco, P. Gentile and C. Noce, Phys. Rev. B, **68**, 054521, 2003.
260. J. Nagamatsu et al., Nature **410**, 63, 2001.
261. P.C. Canfield and G.W. Crabtree, Phys. Today **56**(3), 34, 2003.
262. R. Bel et al., Phys. Rev. Lett. **92**, 177003, 2004.
263. J. Hwany, T. Timusk and G.D. Gu, Nature **692**, 714, 2004.
264. V.L. Ginzburg, *Nobel lecture*, Rev. Mod. Phys. **76**(3), 2004. Chap. 1 in this book.
265. K. Iida et al., Prog. Theor. Phys. Suppl. **153**, 230, 2004.
266. A.I. Buzdin and A.S. Mel'nikov, Phys. Rev. B **67**, 020503(R), 2003.
267. G-H. Gweon et al., Nature **430**, 187, 2004.
268. C.C. Homes et al., Nature **430**, 539, 2004.
269. D. Gudstein and A.R. Chatto, Am. J. Phys. **79**, 85, 2003.
270. T. Löfwander, M. Fogelström, Phys. Rev. B **70**, 024515, 2004.
271. L. Koláček, P. Lipavsky, *Thermopower in superconductors*. cond-mat/0409064; Phys. Rev. B **71**, 092503, 2005.

272. A. Maniv et al., Phys. Rev. Lett. **94**, 247005, 2005.
273. T. Timusk, Phys. World **18**(7), 31, 2005, see also D.N. Basov and T. Timusk, Rev. Mod. Phys. **77**, 721, 2005.
274. A.I. Buzdin, Rev. Mod. Phys. **77**, 935, 2005
275. I.N. Askerzade, Usp. Fiz. Nauk **176**, 1025, 2006 [Phys. Usp. **49**, 1003, 2006].
276. J. Park et al., Nature **440**, 65, 2006.
277. V.L. Belyavsky, Yu.V. Kopaev, Usp. Fiz. Nauk **177**, 565, 2007 [Phys. Usp. **50**, 540, 2007]; J. Supercond. Novel Magnetism **19** (3–5) 251, 2006.
278. E.G. Maksimov and O.V. Dolgov, Usp. Fiz. Nauk **177**, 983, 2007 [Phys. Usp. **50**, 2007].
279. V.Z. Kresin and Y.N. Ovchinnikov, Phys. Rev. B **74**, 024514, 2006. [Phys. Usp. **51**, 2008 (to be published)].
280. A.V. Gurevich, Zh. Eksp. Teor. Fiz. **27**, 195, 1954.

A Few Comments
on Superconductivity Research

In October 2004 and October 2006, the conferences devoted to superconductivity and, in the first place, high-temperature superconductivity (the latter is reflected by the conference name "The Fundamental Problems of High-Temperature Superconductivity") were held near Moscow. These conferences began on 18 October 2004 and 9 October 2006, respectively, and were opened with my comments on the problems discussed. These comments underline the contents of Parts I and II below[1]

3.1 Part I

Superconductivity was discovered in 1911 in Leiden, which was preceded by obtaining liquid helium in 1908. This event can be thought of as the origination of low-temperature physics, although low-temperature studies had certainly begun before that. It would be out of place to dwell here on the history of the development of low-temperature physics and, in particular, the study of superconductivity. Some information on the subject can be found in my paper [1] and Chap. 6 in book [2]. Now I would only like to stress the fact that in an example of the study of superconductivity one can clearly see how radical the changes in science have been over the past century. One can even say that this has all happened within only a century, that is, within a time comparable with the human lifetime and very small, say, compared to the period after the golden age of science in ancient Greece (two or three thousand years ago), to say nothing of the age of homo sapiens (fifty to one hundred thousand years ago).

However, at the beginning of the last century, the pace of the development of physics and science as a whole was incomparable to what we face today.

---

[1] Part I was published in Usp. Fiz. Nauk **175**, 187 (2005) [Phys. Usp. **48**, 173 (2005)]. The whole content of the first conference is reflected in [20]. The second conference revealed in [21]. References [20,21] are included in the list of references to Part I.

Suffice it to say that before 1923, i.e., for 15 long years (on a contemporary scale) liquid helium had been produced in Leiden only, and in that period only a few dozen studies had been performed in the helium temperature range. Within the ten years after the discovery of high-temperature superconductivity (HTSC) in 1986–1987, nearly 50,000 publications were devoted to this subject, that is, 10–15 reports appeared every day.

The style of research also changed rather radically. For example, we have learned only recently [3] that the superconducting transition was first observed quite clearly and definitely by G. Holst, who conducted measurements at the Leiden laboratory. Holst was a qualified physicist (later, the first director of the Philips Research Laboratories and a Professor at Leiden University). Kamerlingh Onnes, however, did not even mention his name in his paper [4], where he reported those measurements. I cannot imagine anything like this happening now in a civilized country (though this is a disputable question, and maybe wrong). At that time, in the early twentieth century, this was obviously the norm in German and congenial universities (the 'Herr Professor', the head of research, could be considered to be the only author of a paper). Such a conclusion seems to me to be well-grounded because, as mentioned in [3], Holst himself had not apparently thought of such a slight by Kamerlingh Onnes as being unjust or unusual. The times were different, and to avoid misunderstanding I will stress that I have no grounds to cast aspersions on Kamerlingh Onnes and his undoubted achievements (which are described in more detail in [1, 2]).

In the 1930s, when I myself began working, low-temperature physics had already occupied an important place in physics in the whole world, and especially in the USSR. In our country, as far as I can judge, this was primarily associated with the activities of L.V. Shubnikov (1901–1937). He graduated from the Leningrad Polytechnical Institute in 1926 and then worked for several years in the Leiden cryogenic laboratory, where he carried out a number of world-famous research works (suffice it to mention the Shubnikov–de Haas effect), and from 1931 up to his untimely demise in 1937 (more precisely, up to his imprisonment a few months before) he was head of the Cryogenic Laboratory at the Kharkov Physicotechnical Institute[2] There he obtained liquid hydrogen in 1931 and liquid helium – for the first time in the USSR – in 1932; liquid helium was only available in a few laboratories in the world (to the best of my knowledge, the second liquefier began operating in 1923 in Toronto and W. Meissner put into operation a helium liquefier in 1925 in Germany). Shubnikov and his students and colleagues accomplished a lot

---

[2] In [5] one reads that L.V. Shubnikov died in 1945. In the Soviet times I happened to hear this false tale, too. However, after the breakup of the USSR some materials were declassified and we learned that Shubnikov, with some of his colleagues, was shot as far back as 1937, soon after his arrest. He was, of course, fully rehabilitated posthumously. L.D. Landau was prosecuted for the same 'case', but he managed to leave for Moscow and was arrested only in 1938. He escaped death literally by a miracle (he was released in 1939; for more details see e.g., [2], paper 10).

within only a few years, and I should especially mention his studies of super-conducting alloys and a factual discovery of type II superconductors (these studies are cited in [6, 7]; see also [5]). I am sure that Shubnikov would have achieved even greater success in science, and one cannot but feel bitterness about his untimely (at the age of only 36!) and quite guiltless death under the ax of Stalin's terror.

Along with the studies by Shubnikov and his school, research work in the field of low-temperature physics began in the 1930s in Moscow in the Institute of Physical Problems at the USSR Academy of Sciences. At that institute, in 1938, and a little earlier, and continuing up to the beginning of the war in 1941 P.L. Kapitza investigated the superfluidity of Helium II [3] and in 1940–1941 L.D. Landau formulated his theory of superfluidity [12]. Both in the prewar years and after the war, many interesting things were done in the USSR in low-temperature physics, but it is hardly pertinent to mention them here (I believe, actually, that it would be most appropriate to devote a special paper or a monograph to this subject).

It seems to me that low-temperature physics occupies in a sense an especially important place in the physical studies that were carried out in the USSR. Suffice it to say, I think, that Soviet (Russian) physicists received six Nobel Prizes in physics, of which half concerned low-temperature physics. In 1962 Landau received a prize 'for the pioneering theories for condensed matter, especially liquid helium.' In 1978 Kapitza received the prize (more precisely, half of the prize) 'for his basic inventions and discoveries in the area of low-temperature physics.' And, finally, in 2003, A.A. Abrikosov, A. Leggett, and I were awarded the prize 'for pioneering contributions to the theory of superconductors and superfluids' [13–15]. The other three prizes were given as follows: in 1958 to I.E. Tamm, I.M. Frank, and P.A. Cherenkov for the discovery and explanation of the Vavilov–Cherenkov effect, in 1964 to N.G. Basov and A.M. Prokhorov (half of the prize) for studies in the field of quantum electronics and, finally, in 2000 to Zh.I. Alferov (part of the prize) for information and communication technology.

Incidentally, I am far from attaching too much importance to Nobel Prizes, as is now done by the mass media. I have always been of this opinion, but before I received the Nobel Prize I had not placed emphasis on it for I would have been suspected of envy, and so on. This is in fact quite clear for one who is aware of the conditions for awarding the prize and knows who was awarded it and who was not (for some more details see paper 21 in [2] and note [16]). At the same time, Nobel Prizes are indicative of the state of a corresponding science in the country. Therefore, I believe that what has been said above

---

[3] The studies of the properties of Helium II began in Leiden as far back as 1911, i.e., the same year that superconductivity was discovered. The landmarks on the long way that led in 1938 to the discovery of superfluidity [8, 9] are mentioned, for example, in [2]. The early stage of the study of superfluidity is described at length in [10] and [11].

shows the particularly high rank of low-temperature physics among the physical research works carried out in our country. I allow myself to accentuate this fact here because from it I shall draw some conclusions about the prospects of the development of physics in Russia. I shall return to this at the end of this chapter.

Since high-temperature superconductors were obtained [17, 18] in 1986–1987 it is naturally their study that has been at the center of attention in the area of low-temperature physics. Almost 20 years have passed since that time, but in spite of numerous studies, the mechanism of HTSC in cuprates is not yet clear enough, and not even at a level of understanding. This situation was elucidated in a number of reviews [19–21]. It was popular for a long time to think of superconductivity in HTSC cuprates as being induced by some exotic mechanisms and, at any rate, as being considerably different from superconductivity in 'conventional' or low-temperature type I superconductors, for which it is well known that the Bardeen, Cooper and Schrieffer (BCS) theory [22] was successfully applied. In this theory, in the simplest case and for 'weak coupling', the critical temperature T_c of superconducting transition is determined by the expression

$$T_c = \theta \exp\left(-\frac{1}{\lambda}\right), \tag{3.1}$$

where $k_B\theta$ is the energy range near a Fermi surface, in which conduction electrons are mutually attracted and λ is the coupling constant which in the weak coupling case is small, i.e.,

$$\lambda \ll 1. \tag{3.2}$$

If the interelectron (quasi-particle) attraction is due to the phonon mechanism (the virtual exchange of phonons), then

$$\theta \sim \theta_D, \tag{3.3}$$

where θ_D is the Debye temperature of metal. The quantity $k_B\theta_D$ is known to be on the order of the maximum energy of the participating phonons. Normally we have $\theta_D < 10^3 K$ with the exception of, say, metallic hydrogen. Hence, it becomes immediately clear why in ordinary metals for electron–phonon coupling with $\lambda < 1$:

$$T_c \lesssim 30 - 40\,\mathrm{K}. \tag{3.4}$$

These simple arguments are well known and have recently been repeated, for example, in [14].

If one assumes materials with $T_c > T_{b\,N_2} = 77.4\,\mathrm{K}$ (the liquid nitrogen boiling temperature at atmospheric pressure) to be high-temperature superconductors, then from (3.4) it is clear that for the weak coupling (3.2) the electron–phonon mechanism will not lead to HTSC.

That is why already in 1964 W.A. Little [23] and then I myself [24] suggested the idea of using the (theoretically possible) attraction between conduction electrons caused by their interaction with bound electrons in the same metal. In vivid language one can speak of the replacement of phonons, i.e., excitations in a crystal (ionic) lattice, by electronic excitons, i.e., excitations in a system of bound electrons. Note that they certainly also include the so-called plasmons and polaritons. Such a mechanism can be called the electron–exciton or simply the exciton mechanism.

The exciton energy $E_{ex} = k_B \theta_{ex}$ in a metal does not exceed the Fermi energy $E_F = k_B \theta_F$ in the order of magnitude. As is known, $E_F \lesssim 10\,\mathrm{eV}$, that is, $\theta_F \lesssim 10^5 \mathrm{K}$. For the exciton mechanism in (3.1)

$$\theta \sim \theta_{ex}. \tag{3.5}$$

Therefore, already at $\theta_{ex} \sim 10^4 \mathrm{K}$ even for weak coupling (3.2) the temperature T_c can assume room values [e.g., at $\theta_{ex} = 10^4 \mathrm{K}$ and $\lambda = 1/3$ we have $T_c = 500\,\mathrm{K}$ according to (3.1)].

Of course, these are now only words because it is not yet clear how, if at all, one can realise an effective exciton mechanism. A group of theoreticians at the Physical Institute of the USSR Academy of Sciences (FIAN) made considerable efforts to investigate the exciton mechanism of HTSC or, more precisely, the whole HTSC problem. The results of these studies are presented in monograph [25]. In line with [24], we underscored in [25] the expedience of using quasi-two-dimensional layered compounds. As is known, this conclusion was later confirmed. Another important result was the establishment, in the known approximation, of metal stability conditions. The point is that one of the main dangers, if not the basic one, is that the lattice and thus the material (crystal) itself may not withstand attempts to raise T_c and might break. In [25] and in references therein it was shown that high T_c values are generally quite agreeable with lattice stability (a somewhat more thorough description of this can be found in [14], see also Chap. 1 in this book). For this reason I shall not give the details here). The fear that high values of the coupling constant λ for the electron–phonon interaction are unrealizable also turned out to be groundless, and the so-called strong coupling with

$$\lambda \gtrsim 1 \tag{3.6}$$

does, in fact, sometimes take place, in particular, in cuprates.

Moreover, the Debye temperature θ_D for cuprates is relatively high. Formula (3.1) is of course invalid in the case of strong coupling, but it nevertheless shows that T_c grows with increasing θ and λ. That is why the values $T_c \lesssim 200\,\mathrm{K}$ can be readily obtained in the case of the electron–phonon mechanism, too (see [19] and papers 6 and 7 in [2]). It is another matter that, certainly, T_c is only one of the characteristics of a superconductor. An account of the electron–phonon interaction alone is probably insufficient to explain all the properties of HTSC cuprates, and obviously one should also allow for the nonsphericity of the Fermi surface and generally the distinction between the conduction-electron spectrum in the crystal and the free-electron spectrum (in BCS theory [22] the conduction electrons or, more precisely, the corresponding quasi-particles were assumed to be free with the exception of their interaction with phonons). In the past, the role of the electron–phonon interaction in the case of HTSC cuprates was frequently considered to be insignificant, in particular, because of the smallness of the isotopic effect (specifically, T_c changes little when the isotope ^{16}O is replaced by the isotope ^{18}O). However, for example, in angle-resolved photoemission spectroscopy the electron excitation spectrum shows a clearly pronounced isotopic effect caused by the electron–phonon interaction in cuprates [26]. Other grounds also exist for such a conclusion, and at the present time there is no doubt that the electron–phonon interaction plays an important, and even perhaps a decisive, role in cuprates.

In any case, if we do not speak of metallic hydrogen (for it $\theta_D \sim 2000$–5000 K), the obtaining of which for practical use is absolutely unrealistic at the present time[4] the possibilities of using the electron–phonon interaction for reaching high, say, room T_c values (that is, the creation of room-temperature superconductors – RTSCs) now clearly seem to be quite limited because the values of the Debye temperature θ_D is in most cases less than several hundred degrees. The same refers to the spin interaction because the Curie temperature θ_C and the Néel temperature θ_N are also typically less than, say, 10^3K (we are speaking, of course, of substances at low pressures). But nonetheless, the attainment of $T_c \sim \theta_C$ or $T_c \sim \theta_N$ and thus the creation of RTSCs on the basis of spin interaction is not excluded. At the same time, as has already been said, the electron–exciton interaction is characterized by the temperature $\theta_{ex} \lesssim \theta_F \lesssim 10^5$K. That is why I think that if room-temperature superconductors can actually be created, it can most likely be possible with the use of the exciton mechanism.

True, I should make an important reservation. Namely, above I have rested upon a BCS-type theory considering the formation of 'pairs' in the s-state resulting from the fact that quasi-free conduction electrons exchange boson type excitations (phonons in the case of the electron–phonon mechanism and excitons for the exciton mechanism). The formation of 'pairs' in p, d, and even in other states is possible if the nonsphericity of the Fermi surface is taken into account. This is well known in the example of the superfluidity of ^3He [15]. The characteristic temperature determining the binding energy of such 'pairs' and, thus, T_c, the same as for θ_{ex}, is lower than or of the order of θ_F. Therefore, even if we forget about the role of phonons and spin interaction, RTSC can in principle be attained not only as a result of the exciton mechanism.

---

[4] As a matter of fact, metallic hydrogen has not yet been obtained even in the laboratory. At the same time, we should point to recent progress in the theoretical study of this substance [27, 28].

Generally, these is no doubt that an unprejudiced approach to the creation of RTSCs is needed. Clearly, at the present time the creation of RTSCs is a typical problem of so-called fundamental science (physics in this case) when we speak of reaching the goal, which is obviously possible in principle, but may be unrealistic. The creation of RTSCs has, of course, great potential regarding their practical use. However, such prospects should not be overestimated, as was the case of the HTSC boom (I an not personally to blame for this). The HTSC materials turned out to be technologically difficult to use, and their application is still rather limited, although some progress has already been made (see, e.g., [29])[5]

It is, of course, not yet time to think about the application of RTSCs, but their creation, I believe, is quite a clear (and, if you like, a very important) task faced by solid state physics.

How can one solve this problem? This is, of course, an enigma. I can only say how I myself would seek such materials if only I could. I remain a follower of the old ideas [24, 25]. Namely, I would seek, or rather create, quasi-two-dimensional layered materials with alternating, at the atomic level, well conducting (metallic) planes-layers and dielectric or, in any case, poorly conducting layers. Such compounds are HTSC cuprates and artificially created layered materials [31]. In addition, one should strive to attain a possibly rich electron exciton spectrum in the system. These excitons must, I repeat, replace phonons, the virtual exchange of which provides, in the case of electron–phonon interaction, an attraction between conduction electrons and their 'pairing'. What has been said is certainly rather vague and not concrete enough. Unfortunately, I was engaged in the study of the exciton theory long ago [32], and now do not follow its development. However, I am aware of its many achievements. The corresponding information should be mobilized and used in attempts to create the materials discussed above. One should also take into consideration some already known and experimentally established regularities that relate the critical temperature T_c to other measurable parameters characterizing superconductors (see, in particular, [33]).

As has already been said, the creation of RTSCs is a clearly outlined and very important goal in solid state physics. Not many other areas of physics can thus boast of the existence of such problems. I shall permit myself to note

---

[5] After the anticipated start in 2008 of the Large Hadron Collider (LHC) at CERN, it is planning the construction of an International Linear Collider (ILC) [30] with a length greater than 30 km. This machine, whose construction will probably begin in 2009, will cost five to seven billion dollars. It will produce two counter-propagating 500-GeV electron or positron beams. I am writing here about it because in the adopted project [30] superconducting magnets are planned to be used at a temperature of 2 K, and these will be conventional superconductors rather than high-temperature superconductors. Hence, the latter cannot in this case compete with conventional (low-temperature) superconductors in spite of the fact that the cooling of the giant machine with liquid helium is much more expensive than with liquid nitrogen.

in conclusion that in Russia close attention to the RTSC problem would be especially justified. This conclusion is grounded, firstly, in view of the tradition of active research in the field of superconductivity in the USSR, which was mentioned at the beginning. Secondly, the corresponding research even at the most up-to-date level requires tens of millions and not billions of dollars that are necessary to create modern ITER, LHC, or ILC type plants [30]. I would like to hope that what has been said above will not pass unnoticed.

In conclusion, I would like to take the opportunity to thank Y.V. Kopaev and E.G. Maksimov for their discussions.

3.2 Part II

It is a known fact, of course, that the studies in the field of superconductivity make up one of the topical directions of research in fundamental physics. This direction is very closely related to so-called applied physics. It was not always so; it suffices to recall that a fairly wide use of superconductivity, discovered in 1911, began only in the 1970s (I obviously mean the creation of strong superconducting magnets). Even high-temperature superconductors (HTSCs), first created in 1987, are now used in engineering even in spite of the well-known difficulties with their treatment. And it is of course clear to everyone that the creation of new 'good' HTSCs that work when cooled in liquid nitrogen, to say nothing of room-temperature superconductors (RTSCs) that do not need any cooling when used at room temperature, will immediately open up broad possibilities for application in practice. The high T_c value is certainly only one requirement on the new substances discussed here, but it is apparently the most hard-hitting. It is scarcely worth writing about these obvious things in more detail. I would only like to emphasize that the study of superconductivity is one of the problems of modern physics which is not only important and interesting, so to say in itself as all fundamental physics, but is also attractive for those who are looking forward to seeing the 'edible fruits' of fundamental science as soon as possible. Incidentally, I myself do not at all belong to those people. More precisely, the optimistic prospects of practical applications are naturally attractive, but should not cause the belittling of the interest in even the most abstract, as it may seem today, problems of science. It is quite another matter that when the means (both financial and other resources, say, manpower) are limited, one should also consider other factors, not only of purely scientific interest. I shall give a concrete example. After the 1918 work of Einstein there arose a problem of generation and reception of gravitational waves (I am sure that the question had also been put forward before, perhaps in the nineteenth century, but I am ignorant of the corresponding literature). Already from Einstein's results it was clear that the corresponding effects are small. Therefore it was only in the 1960s that serious attempts were made to receive gravitational waves of cosmic origin. Unfortunately, the attempts were unsuccessful. It is only now that attempts to discover cosmic gravitational waves, already by other methods (using laser interferometry), are being made in the USA and in Europe. In the USA, the corresponding unit was constructed (more precisely, two units located at a large distance from each other), and the cost of this unit (LIGO) was 500 million dollars. No positive result has as yet been obtained, but there is reason to believe that it will soon be found. However, a more sensitive installation (LIGO II) has now been designed. Incidentally, V.B. Braginskii and his group in the Physics Faculty at Moscow State University take an active part in this work. Reception of gravitational waves will open a new channel of astrophysical information and in future will perhaps be used for communication on the Earth and in the solar system.

So, if I were asked now (and even in the times of the USSR) whether or not I think it necessary to design a LIGO type installation in Russia, I would definitely say no. The reason is clear – one should not undertake anything beyond one's depth. Such a project is more than Russia can afford now. Or, more precisely, the available great means should of course be used to support fundamental science, too, but still with a certain choice because, as we were taught by Kozma Prutkov[6] 'Nobody will embrace the unembraceable.' The study of superconductivity by most advanced methods is precisely the object, of fundamental science as well, to be first investigated. As to the reason for such a choice, I hope it is clear from what has been said above, but it should be added that the foundation of the corresponding modern laboratory, to be mentioned below, will cost only about 15 million dollars. Yes, I mean just 'only' if we compare with the above-mentioned 500 million dollars for LIGO and the billions of dollars spent for the creation of LHC in CERN and, say, the ITER reactor.

In Russia there are unfortunately no such 'superconducting' laboratories, and therefore I applied to President V.V. Putin with a proposition to found it under FIAN. Enclosed in the letter to V.V. Putin, sent on 10 February (2006), was of course the estimate of necessary expenses. I personally have not yet received an answer, although I know from a telephone conversation with the Minister of Education and Science of the Russian Federation, A.A. Fursenko, that he took my proposition positively. But I certainly do not know whether there will be any practical result. If a positive decision is made in the foreseeable future, the foundation of a laboratory will all the same take much time, and some experimental results can be expected in not less than a couple of years.

I believe, by that time the HTSC problem will have been solved in the sense that the pairing mechanism largely responsible for the high T_c values will have been established. Now, a complete discord reigns in the literature in this respect [34], although 20 years have already passed since HTSC were obtained. A complete negation of the role of the electron–phonon interaction (EPI) has

---

[6] Kozma Prutkov, introduced as the author of witty and amusing aphorisms, is a penname of A.K. Tolstoy and the brothers Zhemchuzhnikov.

always seemed to me particularly strange. One of the reasons for this was the absence of the isotope effect in respect of T_c. However, this is explicable also in the presence of the EPI [35]. The main thing is that some characteristics of HTSC samples are known to be strongly affected by the change of isotopes [36]. At the same time, while the smallness of the isotope effect for the EPI can be explained rather easily [35], I do not know how to explain a strong isotope effect in the absence of the EPI. A strong influence that one can see of the replacement of ^{16}O by ^{18}O under tunneling has recently been revealed [37]. Thus, I believe that for the known HTSC a very important role is played by the EPI. The strong electron-phonon coupling in cuprate superconductors is also vividly testified by the experiments performed in [48][7]. Incidentally, this has always been stated by E.G. Maksimov. This circumstance is important from the viewpoint of the creation of new HTSCs with as high T_c values as possible. Indeed, by virtue of what has been said one can hope to increase T_c by changing the structure and thus the lattice of the substance [38, 44]. The discovery in 2001 of MgB_2 superconductivity with $T_c = 40\,K$, which is undoubtedly due to the EPI, is important in this context. It seems to me that the abovementioned remarks [38], made in this connection in respect of the possibility of T_c increase during the search for new structures with the use of the EPI, are very interesting.

The hopes, expressed long ago, for the use of the exciton mechanism [39], i.e., the pairing of conduction electrons due to the interaction with electron excitons (plasmons, polaritons), i.e., Bose excitations in a system of bound electrons, of course, remain in force. In another language, we speak of the influence of bound electrons on T_c (see, in particular, [40, 49–51]). The formation of Cooper pairs is also possible as a result of spin interaction. This has long been known [52] and it has recently been confirmed in [53]. However, according to [54], this mechanism cannot apparently lead to high values of T_c.

The proposals suggested in [45–47] are also worthy of notice in investigating the superconductivity of clusters. With all this, it is of importance that no limitations on T_c values from above are known [39, 51]. An illustration of this circumstance is [41] in which the $T_c \simeq 600\,K$ was obtained for metallic hydrogen at high pressures $p \simeq 20\,Mbar$. Briefly speaking, we do not know any results testifying to the impossibility of creating RTSCs. However, this will undoubtedly be difficult and, perhaps, even beyond the presently avail-

---

[7] This work [48] produces a strong impression – this is the last word in the field of experimental research of HTSCs. Incidentally, if I were an experimenter, I would certainly search for superconductivity in lithium compounds. Lithium itself is superconducting only under pressure (true, even in this case it would be of interest to clarify the magnitude of the isotope effect, i.e., the difference between the behavior of ^6Li and ^7Li). As to its compounds, it cannot be excluded that superconductivity can exist in them even at normal pressure. For the reasons that are clear from my 'Nobel' Autobiography (see Chap. 5 in this book), it would be of particular interest for me to learn about the behavior of ^6LiD (D is deuterium).

able possibilities; thus, the more interesting it is to take on this challenge of nature and try to create RTSCs.

I would like to repeat the assumption which was expressed long ago and then recalled quite recently [42]. I mean that perhaps not all the publications that reported obtaining high T_c values and were then thought of as erroneous were absolutely groundless. Indeed, it is possible that the examined samples had some small inclusions of metastable superconducting phases that decayed with time and, in addition, had been obtained in uncontrollable conditions (an especially well-known example here is CuCl with impurities). Clearly, the suspicious cases concerning this subject should be analyzed again.

Summarizing, I would like to emphasize that in respect of the development of fundamental science in Russia, which is now being frequently written and spoken about, there is every reason and prerequisite to pay special attention to the HTSC and RTSC problems. I spoke about it at the previous conference on superconductivity, which was held here in 2004 (see [43]).

References

Part I

1. V.L. Ginzburg, *Sverkhprovodimost'*: *Fiz., Khim., Tekh.* **5**, 1, 1992. [*Supercon-ductivity: Phys. Chem. Technol.* **5**, 1, 1992].
2. V.L. Ginzburg, *O Nauke, o Sebe i o Drugikh* (About Science, Myself, and Others) (Moscow: Fizmatlit, 2003, 2004) [Translated into English (Bristol: IOP Publ., 2005)].
3. J. de Nobel, Phys. Today **49**(9), 40, 1996.
4. H. Kamerlingh Onnes, Commun. Phys. Lab. Univ. Leiden, 124 c, 1911[8]
5. Yu.A. Khramov, *Fiziki (Biograficheskiĭ Spravochnik)* (Physicists' (Biographi-cal Reference Book)) (Ed. A.I. Akhiezer) 2nd edn. (Moscow: Nauka, 1983).
6. V.R. Karasik (Ed.), *Sverkhprovodimost'. Bibliograficheskiĭ Ukazatel'. 1911–1970* (Superconductivity. Bibliographical Index. 1911–1970) (Moscow: Nauka, 1975).
7. V.L. Ginzburg, *Sverkhprovodimost'* (Superconductivity) (Moscow–Leningrad: Izd. AN SSSR, 1946).
8. P.L. Kapitza, Nature **141**, 74, 1938.
9. F. Allen, A.D. Misener, Nature **141**, 75, 1938; Proc. R. Soc. London Ser. A **172**, 467, 1939.
10. P.L. Kapitza, Zh. Eksp. Teor. Fiz. **11**, 581, 1941; J. Phys. USSR **4**, 181; **5**, 59, 1941.
11. W.H. Keesom, Helium (Amsterdam: Elsevier, 1942). [Translated into Russian (Moscow: IL, 1949)].
12. L.D. Landau, Zh. Eksp. Teor. Fiz. **11**, 592, 1941; J.Phys. USSR **5**, 71, 1941.
13. A.A. Abrikosov, 'Nobel lecture' Usp. Fiz. Nauk **174**, 1234, 2004; Rev. Mod. Phys. **76**, 975, 2004.

---

[8] This communication is also published as a supplement in [1].

14. V.L. Ginzburg, *Nobel lecture*. Usp. Fiz. Nauk **174**, 1240, 2004. [Phys. Usp. **47**, 1155, 2004)]; Rev. Mod. Phys. **76**, 981, 2004. Chapter 1 in this book.

15. A.G. Leggett, *Nobel Lecture*. Usp. Fiz. Nauk **174**, 1256, 2004; Rev. Mod. Phys. **76**, 999, 2004.

16. V.L. Ginzburg, "Neskol'ko zamechaniĭ o Nobelevskikh premiyakh" ("Some remarks on Nobel prizes") Universum (3) 40, 2004.

17. J.G. Bednorz, K.A. Müller, Z. Phys. B **64**, 189, 1986.

18. M.K. Wu et al., Phys. Rev. Lett. **58**, 908, 1987.

19. E.G. Maksimov, Usp. Fiz. Nauk **170**, 1033, 2000; **174**, 1026, 2004. [Phys. Usp. **43**, 965, 2000; **47**, 957, 2004].

20. V.I. Belyavskii, Yu.V. Kopaev, Usp. Fiz. Nauk **175**, 191, 2005. [Phys. Usp. **48**, 177, 2005].

21. Proc. of the 2nd Conference "Fundamental Problems of High-Temperature Superconductors FPS'06", Moscow, Zvenigorod, 9 October 2006.

22. J. Bardeen, L.N. Cooper, and J.R. Schrieffer, Phys. Rev. **108**, 1175, 1957.

23. W.A. Little, Phys. Rev. **134**, A1416, 1964.

24. V.L. Ginzburg, Zh. Eksp. Teor. Fiz. **47**, 2318, 1964. [Sov. Phys. JETP **20**, 1549, 1965]; Phys. Lett. **13**, 101, 1964.

25. V.L. Ginzburg, D.A. Kirzhnits (Eds.) *Problema Vysokotemperaturnoĭ Sverkh-provodimosti* (High-Temperature Superconductivity) (Moscow: Nauka, 1977) [Translated into English (New York: Consultants Bureau, 1982)].

26. G-H. Gweon et al., Nature **430**, 187, 2004.

27. E. Babaev, A. Sudbø, N.W. Ashcroft, Nature **431**, 666, 2004.

28. S.A. Bonev et al., Nature **431**, 669, 2004.

29. A.M. Campbell, Phys. World **17** (8), 37, 2004.

30. T. Feder, Phys. Today **57** (10), 34, 2004.

31. I. Bozovic, J.K. Eckstein, in *Physical Properties of High Temperature Super-conductors V* (Ed. D.M. Ginsberg) (Singapore: World Scientific, 1996); I. Bozovic, IEEE Trans. Appl. Supercond. **11**, 2686, 2001; I. Bozovic et al. Phys. Rev. Lett. **89**, 107001, 2002; Nature **422**, 873, 2003.

32. V.M. Agranovich, V.L. Ginzburg, *Kristallooptika s Uchetom Prostranstvennoĭ Dispersii i Teoriya Eksitonov* (Crystal Optics with Spatial Dispersion, and Excitons) 2nd edn. (Moscow: Nauka, 1979) [Translated into English (Berlin: Springer, 1984)]. (Now the new edition is prepared at Springer.)

33. C.C. Homes et al., Nature **430**, 539, 2004; J. Zaanen, Nature **430**, 512, 2004.

Part II

34. D.A. Bonn, Nature Phys. **2**, 159, 2006 (the discussion on the nature of HTSC is published in this issue, in which A. Leggett, D. Pines, and others participated).

35. A.E. Karakozov and E.G. Maksimov, Zh. Eksp. Teor. Fiz. **74**, 681, 1978 [Sov. Phys. JETP **47**, 358, 1978].

36. G.A. Gweon et al., Nature **430**, 187, 2004.

37. J. Lee et al., Nature **442**, 546, 2006.

38. W.E. Pickett, cond-mat/0603482, cond-mat/0603428; J. Supercond. and Novel Magn. **19**, 291, 2006.

39. *Problema vysokotemperaturnoi sverkhprovodimosti* (*Problem of high-tempe-rature superconductivity*). Ed. V.L. Ginzburg and D.A. Kirzhnits. Moscow, Nauka, 1977 [High-Temperature Superconductivity (Eds. V.L. Ginzburg, D.A. Kirzhnitz) (New York: Consultants Bureau, 1982)].

40. W.A. Little, *Physica C in press* (Talk at the 8th International HTSC Conference, Dresden, July 2006).
41. E.G. Maksimov and S.Yu. Savrasov, Solid State Commun. **119**, 569, 2001.
42. T.H. Geball, J. Supercond. and Novel Magn. **19**, 261, 2006; see also [44].
43. V.L. Ginzburg, *A few remarks on the study of superconductivity*. Usp. Fiz. Nauk **175**, 187, 2005. [Phys.-Usp. **48**, 173, 2005] This note is placed above as Part I of the present chapter.
44. M.L. Cohen, J. Supercond. and Novel Magn. **19**, 283, 2006.
45. Y.N. Ovchinnikov and V.Z. Kresin, Eur. Phys. J. B **45**, 5, 2005; **47**, 333, 2005.
46. V.Z. Kresin and Y.N. Ovchinnikov, Phys. Rev. B **74**, 024514, 2006.
47. V.Z. Kresin and Y.N. Ovchinnikov, Usp. Fiz. Nauk **178**, 2008 [Phys. Usp. **51**, 2008 (in press)].
48. I. Bozović, Usp. Fiz. Nauk **178**, 179, 2008 [Phys. Usp. **51**(2), 2008 (in press)].
49. W.A. Little, J. Supercond. Nov. Magn. **19**, 443, 2006.
50. F.C. Niestemski et al., Nature **450**, 1058, 2007.
51. E.G. Maksimov and O.V. Dolgov, Usp. Fiz. Nauk **177**, 983, 2007 [Phys. Usp. **50**, 933, 2007].
52. A.I. Akhiezer and I.Ya. Pomeranchuk, Zh. Eksp. Teor. Fiz. **36**, 859, 1959 [Sov. Phys. JETP **36**, 605, 1959].
53. P. Monthoux, D. Pines and G.G. Lonzarich, Nature **450**, 1177, 2007.
54. E.G. Maksimov, Usp. Fiz. Nauk **178**, 175, 2008 [Phys. Usp. **51**(2), 2008 (in press)].

4

On the Theory of Superconductivity[1]

by V.L. Ginzburg and L.D. Landau

The existing phenomenological theory of superconductivity is unsatisfactory, since it does not allow us to determine the surface tension at the boundary between the normal and the superconducting phases, and does not allow for the possibility to describe correctly the destruction of superconductivity by a magnetic field or current. In the present paper, a theory is constructed which is free from these faults. We find equations for the Ψ-function of the 'superconducting electrons' which we introduced and for the vector potential. We have solved these equations for the one-dimensional case (a superconducting half-space and flat plates).

The theory makes it possible to express the surface tension in terms of the critical magnetic field and the penetration depth of the magnetic field in superconductors. The penetration depth depends in a strong field on the field strength and this effect will be especially evident in the case of small size superconductors. The destruction of superconductivity in thin plates by a magnetic field is through a second-order phase transition, and it only becomes a first-order transition starting with plates of a thickness more than a certain critical thickness. While the critical external magnetic field increases with the decreasing thickness of the plates, the critical current for destroying the superconductivity of plates decreases with decreasing thickness.

4.1 Introduction

It is well known that there exists at present no properly developed microscopic theory of superconductivity. At the same time there is a fairly widespread view that the phenomenological theory of superconductivity is in a much more satisfactory state and is reliably based on the equation of F. London and H. London [1, 2]:

---

[1] This paper was first published in Russian in Zh. Eksp. Teor. Fiz. (ZhETF) **20**, 1064 (1950).

$$\operatorname{rot} \Lambda \boldsymbol{j}_s = -\frac{1}{c} \boldsymbol{H}, \tag{4.1}$$

where Λ is a quantity depending only on temperature, $\boldsymbol{j}_s$ is the supercurrent density, c is the velocity of light and $\boldsymbol{H}$ is the magnetic field strength, here identical with the magnetic induction.

Equation (4.1), in combination with Maxwell's equation, $\operatorname{rot} \boldsymbol{H} = (4\pi/c)\boldsymbol{j}_s$, and the equations div $\boldsymbol{H} = 0$ and div $\boldsymbol{j}_s = 0$, leads under stationary conditions to the equations

$$\nabla^2 \boldsymbol{H} - \frac{\boldsymbol{H}}{\delta^2} = 0 \quad \text{and} \quad \nabla^2 \boldsymbol{j}_s - \frac{\boldsymbol{j}_s}{\delta^2} = 0, \quad \text{where} \quad \delta^2 = \frac{\Lambda c^2}{4\pi}. \tag{4.2}$$

For a plane boundary between the superconductor and vacuum or a non-superconductor, these equations have solutions

$$H = H_0 \exp\left(-\frac{z}{\delta}\right) \quad \text{and} \quad j_s = \frac{c}{4\pi\delta} H, \tag{4.3}$$

in which the external field H_0 is taken as parallel to the boundary, which is normal to the z-axis. For a film of thickness $2d$ in a parallel field we get

$$H = H_0 \frac{\cosh(z/\delta)}{\cosh(d/\delta)}, \quad j_s = -\frac{cH_0}{4\pi\delta} \frac{\sinh(z/\delta)}{\cosh(d/\delta)}, \tag{4.4}$$

if $z = 0$ at the centre of the film.

For a superconductor of arbitrary shape it follows from (4.2) that the field penetrates only to a depth of the order of δ, which is, according to experimental data, about 10^{-5}cm. Qualitatively, this result is of course in agreement with the fact that a magnetic field does not penetrate into the body of a superconductor; quantitatively, however, there is no certainty that equations (4.1)–(4.4) are always correct. Moreover, this theory throws no light on the question of the surface energy at a boundary between superconducting and normal phases of the same metal, and also leads to a contradiction with the experiment concerning the destruction of superconductivity of a thin film by a magnetic field.

The thermodynamic treatment of the transition of a film of thickness $2d$ from the superconducting to the normal state leads [2, 3] to the following expression for the critical field, H_c:

$$\left(\frac{H_c}{H_{cb}}\right)^2 = \left(1 - \frac{\delta}{d}\tanh\frac{d}{\delta}\right)^{-1} \tag{4.5}$$

in which H_{cb} is the critical field of the bulk material. This expression is not in agreement with the experiment. Thus, if at a given temperature the constant δ is determined from measured values of $(H_c/H_{cb})^2$ for various values of d, according to (4.5), then this "constant" δ depends markedly on d; for example, if $T = 4$K then for $d = 0.3 \times 10^{-5}$cm, $\delta = 3.4 \times 10^{-5}$cm, while for $d = 1.2 \times 10^{-5}$cm, $\delta = 2 \times 10^{-5}$cm.

It has been pointed out [3] that the position may be improved by taking into account the difference of the surface energy at the boundary of the metal with a vacuum according as the metal is in the superconducting or the normal state; the difference of surface energies, $\sigma_s - \sigma_n$, introduced for this purpose, must be of the order of $\delta H_{cb}^2/8\pi$. Now the surface energy is usually equal to the bulk free energy per unit volume times a length of the order of atomic dimensions. Thus, here, where the difference of free energies is $H^2/8\pi$, one might expect $\sigma_s - \sigma_n$ to be of the order of 10^{-7} to 10^{-8} times $H_{cb}^2/8\pi$ and not $10^{-5}H_{cb}^2/8\pi$. An even more contradictory situation arises at the boundary separating the normal and superconducting phases of the metal; the surface energy connected with the field and supercurrent here as predicted from the solution of (4.3), is equal [3,5] to $-\delta H_{cb}^2/8\pi$, i.e., is negative. Thus, in order to obtain the observed positive surface energy σ_{ns}, it is necessary to introduce a surface energy, σ'_{ns}, of non-magnetic origin, which is given by the equation

$$\sigma'_{ns} = \sigma_{ns} + \frac{\delta H^2}{8\pi}$$

and which is greater than $\delta H_{cb}^2/8\pi$. There is no justification for introducing such a relatively enormous energy σ'_{ns} not connected with the field distribution. On the contrary, one would expect any rational theory of superconductivity to lead automatically to an expression for σ_{ns} in terms of the ordinary parameters characterising the superconductor.

The theory based on (4.1), even with the additional surface energy, does not enable the destruction of superconductivity in thin films by a current [6] to be considered, since this problem is not of a thermodynamic nature.

The aim of the present work is the construction of a theory free from these defects. Incidentally, as we shall see, the theory leads also to a number of new qualitative conclusions, which may be checked experimentally.

4.2 Basic Equations

In the absence of a magnetic field, the transition into the superconducting state at the critical temperature T_c is a phase transition of the second kind. In the general theory of such transitions [7] there always enters some parameter η which differs from zero in the ordered phase and which equals zero in the disordered phase. For example, in ferroelectrics, the spontaneous polarisation plays the role of η and in ferromagnetics the spontaneous magnetisation [8]. In the phenomenon of superconductivity, in which it is the superconducting phase that is ordered, we shall use Ψ to denote this characteristic parameter. For temperatures above T_c, $\Psi = 0$ in the state of thermodynamic equilibrium, while for temperatures below T_c, $\Psi \neq 0$. We shall start from the idea that Ψ represents some 'effective' wave function of the 'superconducting electrons.' Consequently, Ψ may be precisely determined only apart from a phase constant. Thus, all the observable quantities must depend on Ψ and Ψ^* in such a way that they are unchanged when Ψ is multiplied by a constant of the type $\exp(i a)$. We may note also that since the quantum mechanical connection between Ψ and the observable quantities has not yet been determined, we may normalise Ψ in an arbitrary manner. We shall see below how we must carry out this normalisation in such a way that $|\Psi|^2$ shall equal the concentration, n_s, of 'superconducting electrons' introduced in the usual way.

Consider firstly a uniform superconductor in the absence of a magnetic field, and suppose that Ψ is independent of position. The free energy of the superconductor is then in accordance with the general theory of second-order phase transitions, dependent only on $|\Psi|^2$ and may be expanded in series form in the neighbourhood of T_c. Thus, near T_c, we may write for the free energy F_{s0}:

$$F_{s0} = F_{n0} + \alpha|\Psi|^2 = \frac{\beta}{2}|\Psi|^4. \tag{4.6}$$

In equilibrium $\partial F_{s0}/\partial|\Psi|^2 = 0$, $\partial^2 F_{s0}/\partial^2|\Psi|^2 > 0$, and, in addition, we must have that $|\Psi|^2 = 0$ for $T \geq T_c$ and $|\Psi|^2 > 0$ for $T > T_c$. It follows, therefore, that $\alpha_c = 0$, $\beta_c > 0$, and for $T < T_c$, $\alpha < 0$. Thus, in equilibrium, for $T \leq T_c$

$$|\Psi|^2 \equiv |\Psi_\infty|^2 = -\frac{\alpha}{\beta} = \frac{T_c - T}{\beta_c}\left(\frac{\mathrm{d}\alpha}{\mathrm{d}T}\right)_c,$$

and

$$F_{s0} = F_{n0} - \frac{\alpha^2}{\alpha\beta} = F_{n0} - \frac{(T_c - T)^2}{2\beta_c}\left(\frac{\mathrm{d}\alpha}{\mathrm{d}T}\right)_c^2, \tag{4.7}$$

in which it is taken into account that, within the limits of validity of the expansion (4.6), $\alpha(T) = (\mathrm{d}\alpha/\mathrm{d}T)_c(T_c - T)$ and $\beta(T) = \beta_c$; the choice of the subscript ∞ for Ψ is determined by considerations of convenience which will become evident from what follows. The quantity F_{n0} in (4.6) and (4.7) is evidently the free energy of the normal phase. Well-known thermodynamic considerations show (see also below) that $F_{s0} - F_{n0} = H_{cb}^2/8\pi$, where H_{cb} is the critical magnetic field for a bulk specimen, and the free energies, as everywhere in this paper, relate to unit volume. Thus, from (4.7)

$$H_{cb}^2 = \frac{4\pi\alpha^2}{\beta} = \frac{4\pi(T_c - T)^2}{\beta_c}\left(\frac{\mathrm{d}\alpha}{\mathrm{d}T}\right)_c^2. \tag{4.8}$$

The form of this expression is well known to be completely confirmed by an experiment, which therefore provides a foundation for the assumptions made above.

Consider now a superconductor in a time-independent magnetic field. In order to obtain the density of the total free energy F_{sH}, it is now necessary to add to F_{s0} the field energy $H^2/8\pi$ and the energy connected with the possible appearance of a gradient of Ψ in the presence of a field. This last energy, at least for small values of $|\nabla\Psi|^2$, can as a result of a series expansion with respect to $|\nabla\Psi|^2$ be expressed in the form const $|\nabla\Psi|^2$, i.e., it looks like the density of kinetic energy in quantum mechanics. Thus, we shall write the corresponding expression in the form

$$\left(\frac{\hbar^2}{2m}\right)|\nabla\Psi|^2 = \frac{1}{2m}|-i\hbar\nabla\Psi|^2$$

in which $\hbar = 1.05 \times 10^{-27}$ is Dirac's constant and m is a certain coefficient. We have not, however, taken into account as yet the interaction between the magnetic field and the current connected with the presence of $\nabla\Psi$. In view of what has been said, and the requirement that the whole scheme shall be gauge-invariant, we must allow for the influence of the field by making the usual change of $i\hbar$ grad to $[-i\hbar\nabla-(e/c)\mathbf{A}]$, where $\mathbf{A}$ is the vector potential of the field and e is a charge, which there is no reason to consider as different from the electronic charge. Thus, the energy density connected with the presence of $\nabla\Psi$ and the field H takes the form

$$\frac{H^2}{8\pi} + \frac{1}{2m}\left|-i\hbar\nabla\Psi - \frac{e}{c}\mathbf{A}\Psi\right|^2 .$$

Consequently,

$$F_{sH} = F_{sn} + \frac{H^2}{8\pi} + \frac{1}{2m}\left|-i\hbar\nabla\Psi - \frac{e}{c}\mathbf{A}\Psi\right|^2 . \tag{4.9}$$

The equation for Ψ may now be found from the requirement that the total free energy of the body, $\int F_{sH}\,dV$, shall be as small as possible. Thus, varying with respect to Ψ^*, we find that,

$$\frac{1}{2m}\left(-i\hbar\nabla - \frac{e}{c}\mathbf{A}\right)^2\Psi + \frac{\partial F_{s0}}{\partial\Psi^*} = 0 \tag{4.10}$$

and, moreover, at the boundary of the superconductor, in view of the arbitrariness of the variation $\delta\Psi^*$, the following condition must hold

$$\mathbf{n}\cdot\left[-i\hbar\nabla\Psi - \frac{e}{c}\mathbf{A}\Psi\right] = 0 \tag{4.11}$$

where $\mathbf{n}$ is the unit vector normal to the boundary.

The condition (4.11) is obtained if no supplementary requirements are imposed on Ψ (natural boundary conditions); if, however, it is demanded from the start that at the boundary with a vacuum $\Psi = 0$, then (4.11) is not obtained. But, the condition $\Psi = 0$ or const is not admissible in the present scheme, since then there would be no solution to the problem of the superconducting plate except for particular values of the thickness $2d$. We therefore impose no further conditions on Ψ at the boundary with a vacuum, and are thus led to (4.11). At first sight, this result may appear unacceptable, since it is natural to demand that the wave function at the boundary of a metal should vanish. The essence of the matter, however, lies in the fact that the Ψ-function introduced above is in no way a true wave function of the electrons in the metal, but is a certain average quantity.

We may suppose that our function $\Psi(r)$ is directly connected with the density-matrix $\rho(r, r') = \int \Psi^*(r, r'_i)\Psi(r, r'_i)\,dr'_i$, where $\Psi(r, r'_i)$ is the true wavefunction of the electrons in the metal, depending on the coordinates of all electrons, r_i $(i = 1, 2, \ldots, N)$; the r'_i are the coordinates of all the electrons except the one considered, whose coordinates at two points are taken as r and r'. It might be thought that when $|r - r'| \to \infty$, $\rho = 0$ for a non-superconducting body having no long-range order, while in the superconducting state $\rho(|r - r'|) \to \infty) \to \rho_0 \neq 0$. It is reasonable to suppose now that the density-matrix is connected with our Ψ-function by the relation $\rho(r, r') = \Psi^*(r)\Psi(r')$.

So far as the equation for A is concerned, if we assume that div $A = 0$ and vary the free energy with respect to A, we obtain the usual expression

$$\nabla^2 A = -\frac{4\pi}{c}j = \frac{2\pi i e \hbar}{2mc}(\Psi^*\nabla\Psi - \Psi\nabla\Psi^*) + \frac{4\pi e^2}{mc^2}|\Psi|^2 A, \qquad (4.12)$$

in which the right-hand side contains the expression for the supercurrent

$$j = -\frac{ie\hbar}{2m}(\Psi^*\nabla\Psi - \Psi\nabla\Psi^*) - \frac{e^2}{mc}\Psi^*\Psi A.$$

It should be noticed that an expression analogous to (4.11) is obtained for the quantity in brackets, from which it is evident that at the boundary $(j \cdot n) = 0$, as required. The solution of the problem of the distribution of field and current in a superconductor is now reduced to an appropriate integration of (4.10) and (4.12).

We shall examine below only the one-dimensional problem, with the z-axis normal to the boundary separating the superconducting phase $(z > 0)$ from the normal phase or vacuum; we shall take the field H as directed along the y-axis and the current j and vector potential A along the x-axis (thus $H_y = dA_x/dz$, or simply $H = dA/dz$). In the one-dimensional solution it is natural to consider $|\Psi|^2$ as dependent only z, so that $\Psi = \exp[i\varphi(x, y)]\Psi(z)$. However, bearing in mind the gauge-invariance of the equations, we may by a suitable choice of A arrange that $\Psi = \Psi(z)$ and hence $j = -e^2/mc|\Psi|^2 A$ (from the conditions that div $j = dj_z/dz = 0$ and $(j \cdot n) = 0$ it follows that $j_z = 0$). Moreover, the equations do not now contain the imaginary i (since $(A \cdot \nabla\Psi) = (Aj \cdot (d\Psi/dz)k) = 0$), and we may therefore consider Ψ as real. Consequently, (4.10) and (4.12) take the form

$$\frac{d^2\Psi}{dz^2} + \frac{2m}{\hbar^2}|\alpha|\left(1 - \frac{e^2}{2mc^2|al|}A^2\right)\Psi - \frac{2m}{\hbar^2}\beta\Psi^3 = 0,$$

$$\frac{d^2A}{dz^2} - \frac{4\pi e^2}{mc^2}\Psi^2 A = 0 \qquad (4.13)$$

in which (4.6) has been used, with the additional fact that $\alpha < 0$.

Let us now determine the surface energy at a plane boundary between the normal and superconducting phases. In the normal phase, the total free energy, including the field energy, is $F_{n0} + H_{cb}^2/8\pi$. In the region where $\Psi \neq 0$, and there is superconductivity, the energy density is F_{sH} (4.9), and, in addition, we must take account of the energy density due to the 'magnetization' of a superconductor in a field parallel to the boundary, with the nonsuperconducting phase, in the form

$$-MH_{cb} = -\frac{H(z) - H_{cb}}{4\pi} H_{cb},$$

where M plays the role of the magnetization. Thus, the surface energy may be written,

$$\sigma_{ns} = \int \left(F_{sH}(z) - \frac{H(z)H_{cb}}{4\pi} + \frac{H_{cb}^2}{4\pi} - F_{n0} - \frac{H_{cb}^2}{8\pi} \right) dz, \qquad (4.14)$$

in which the integration is extended over the transition layer between the phases (the z-axis is normal to this layer). It is readily verified that the integrand vanishes at great distances from the transition layer, for in the superconducting phase $H = 0$ and $F_{sH} = F_{s0} = F_{n0} - \alpha^2/2\beta$ [see (4.7)], while in the normal phase $\Psi = 0$, $F_{sH} = F_{n0} + H_{cb}^2/8\pi$ and $H_0 = H_{cb}$. From (4.7)–(4.9)

$$\sigma_{ns} = \int \left\{ \alpha\Psi^2 + \frac{\beta\Psi^4}{2} + \frac{\alpha^2}{2\beta} + \frac{\hbar^2}{2m}\left(\frac{d\Psi}{dz}\right)^2 + \frac{e^2}{2mc^2}A^2\Psi^2 + \frac{H^2}{8\pi} - \frac{H_{cb}H}{4\pi} \right\} dz.$$
$$(4.15)$$

From the minimum condition for σ_{ns}, which is the free energy per unit area, we may, of course, obtain both the first of equations (4.13), by variation of (4.15) with respect to Ψ, and the second of equations (4.13), by variation with respect to A.

At the boundary of a superconductor with a vacuum in the one-dimensional case, (4.11) assumes the form

$$\frac{d\Psi}{dz} = 0. \qquad (4.16)$$

We shall now introduce the following parameters, H_{cb}, δ_0 and $\varkappa$ and, in addition, the new variables, z', Ψ', A' and H':

$$z' = \frac{z}{\delta_0}, \quad \Psi'^2 = \frac{\Psi^2}{\Psi_\infty^2} = \frac{\Psi^2}{|\alpha|/\beta}, \quad A' = \sqrt{\frac{e^2}{2mc^2|\alpha|}}\, A = \frac{A}{\sqrt{2}\,H_{cb}\delta_0},$$

$$H' = \frac{dA'}{dz'} = \frac{1}{\sqrt{2}}\frac{H}{H_{cb}}, \quad \delta_0^2 = \frac{mc^2\beta}{4\pi e^2|\alpha|} = \frac{mc^2}{4\pi e^2 \Psi_\infty^2},$$

$$H_{cb}^2 = \frac{4\pi\alpha^2}{\beta}, \quad \varkappa^2 = \frac{1}{2\pi}\left(\frac{mc}{e\hbar}\right)^2 \beta = \frac{2e^2}{\hbar^2 c^2}H_{cb}^2\delta_0^4. \qquad (4.17)$$

Equations (4.13) now take the form

$$\frac{d^2\Psi}{dz^2} = \varkappa^2[-(1-A^2)\Psi + \Psi^3], \quad \frac{d^2A}{dz^2} = \Psi^2 A. \tag{4.18}$$

The primes have been omitted from these equations, since in what follows, unless we explicitly state the contrary, only the new variables will be used. With these variables, (4.15) must be written in the form

$$\sigma_{ns} = \frac{H_{cb}^2}{4\pi}\delta_0 \int \left\{ \frac{1}{2} - (1-A^2)\Psi^2 + \frac{1}{2}\Psi^4 + \frac{1}{\varkappa^2}\left(\frac{d\Psi}{dz}\right)^2 \right.$$

$$\left. + \left(\frac{dA}{dz}\right)^2 - 2\left(\frac{dA}{dz}\right)_c\left(\frac{dA}{dz}\right) \right\} dz. \tag{4.19}$$

If $\varkappa = 0$, then from (4.18) and (4.16) $\Psi^2 = n_s = $ const, and our equations go over into (4.2) with $\delta^2 = \delta_0^2 = mc^2/4\pi e^2 n_s$ [compare (4.2) with the second of equations (4.13)]. This result is true in general; if we put $\nabla\Psi = 0$ in (4.12) it becomes equivalent to (4.2), or, more directly, $\boldsymbol{j} = -(e^2/mc)|\Psi|^2\boldsymbol{A}$, which leads to (4.1). Although for $\varkappa = 0$ our scheme becomes formally identical with the usual theory, it is substantially different, even in this limiting case. For in (4.1) and (4.2) the parameter $\Lambda = 4\pi\delta^2/c^2 = m/n_s e^2$ is a constant, independent of the field, at a given temperature, while in our theory, even for $\varkappa = 0$, the value of Ψ^2 which is the same as n_s and which determines, as in (4.20), the value of δ, is such as to minimise the free energy, and this results in a variation of the penetration depth δ with H in superconductors of small dimensions.

From the limiting case, $\varkappa = 0$, and from the following discussion, it is clear that the experimentally determined quantity is the parameter $\delta_0^2 = mc^2/4\pi e^2\Psi_\infty^2$, δ_0 being the penetration depth for a weak field into a bulk superconductor. It is just this quantity which enters also into the expression for the dielectric constant $\varepsilon = \varepsilon_0 - 4\pi e^2\Psi_\infty^2/m\omega^2$ of a superconductor in an alternating field of not too high a frequency ω (ε_0 is a certain constant contribution to ε from all particles other than 'superconducting electrons'). The parameter $\Psi_\infty^2 = n_s$, which evidently corresponds to the concentration of 'superconductive electrons', does not appear as a measurable quantity, resembling in this way the number of free electrons in the ordinary quantum theory of metals. Thus, in both expressions, we may talk only of the effective number of electrons which may be determined from the values of ε or δ_0^2 by attributing to m the value appropriate to a free electron. Proceeding in this way, we relate the concentration of 'superconducting electrons' $n_s = \Psi_\infty^2$ with the observable quantity δ_0 (putting $e = 4.8\times10^{-10}$e.s.u., $m = 9.1\times10^{-28}$g) by

$$\delta_0^2 = \frac{mc^2\beta}{4\pi e^2|\alpha|} = 2.84 \times 10^{11}\frac{\beta}{|\alpha|} = \frac{2.84 \times 10^{11}}{\Psi_\infty^2} \text{ cm}^2. \tag{4.20}$$

From (4.20) and from measurements of the critical field $H_{cb} = \sqrt{4\pi\alpha^2/\beta}$ we may determine α and β. Besides H_{cb} and δ_0 (or α and β), there also enters into the theory the dimensionless parameter $\varkappa$

$$\varkappa^2 = \frac{2e^2}{\hbar^2 c^2} H_{cb}^2 \delta_0^4. \tag{4.21}$$

which, with $e = 4.8 \times 10^{-10}$ e.s.u., becomes

$$\varkappa^2 = 4.64 \times 10^{14} H_{cb}^2 \delta_0^4, \tag{4.22}$$

where δ_0 is measured in cm and H_{cb} in gauss. From the experimental data discussed in Sect. ?? it follows that for mercury

$$\varkappa^2 \approx 0.027; \quad \varkappa \approx 0.165; \quad \sqrt{\varkappa} \approx 0.406. \tag{4.23}$$

4.3 The Superconducting Half-Space

We shall consider first the case of a superconducting half-space bounded by a vacuum (superconducting for $z > 0$, boundary at $z = 0$). The solution will, of course, refer also to a sufficiently thick plate whose half thickness $d \gg 1$ (or in the usual units $d \gg \delta_0$). For $z = 0$, $H = H_0$, and for $z = \infty$, $H = A = 0$ (the present choice of $A(\infty) = 0$ is perfectly natural, and moreover, possible). Furthermore, for $z = \infty$ we are dealing with a superconductor in the absence of a field and far from any boundaries, and consequently (4.7) must apply, i.e., in the new variables $\Psi_\infty^2 = 1$, $d\Psi/dz = 0$. Thus, for $z = \infty$,

$$\Psi_\infty^2 = 1, \quad \frac{d\Psi}{dz} = H = A = 0. \tag{4.24}$$

This solution naturally satisfies (4.18). As regards the boundary with the vacuum at $z = 0$, condition (4.16) must be satisfied there; substituting (4.18) into (4.24) we see that in the absence of a magnetic field, the presence of the boundary has no influence on the function Ψ, which, therefore, has the same value everywhere:

$$\Psi^2 = \Psi_\infty^2 = 1 \quad \text{if} \quad H \equiv A \equiv 0. \tag{4.25}$$

In the presence of a magnetic field (4.25), of course, does not apply, and we must integrate (4.18) with the boundary conditions (4.24) for $z = \infty$ and the conditions

$$H = \frac{dA}{dz} = H_0, \quad \frac{d\Psi}{dz} = 0 \quad \text{for} \quad z = 0. \tag{4.26}$$

The values of A_0 and Ψ_0 are not known beforehand.

The equations (4.18), unfortunately, cannot be integrated exactly, and we can indicate only one of their integrals,

$$(1 - A^2)\Psi^2 - \frac{1}{2}\Psi^4 + \left(\frac{dA}{dz}\right)^2 + \frac{1}{\varkappa^2}\left(\frac{d\Psi}{dz}\right)^2 = \text{const.} \tag{4.27}$$

For the case which interests us, the const $= 1/2$ because of (4.24), and thus

$$H^2 = \left(\frac{\mathrm{d}A}{\mathrm{d}z}\right)^2 = \frac{1}{2} - \frac{1}{\varkappa^2}\left(\frac{\mathrm{d}\Psi}{\mathrm{d}z}\right)^2 - (1 - A^2)\Psi^2 + \frac{1}{2}\Psi^4. \tag{4.28}$$

Turning instead to the approximate solution of (4.18), we now give the solution valid for small values of $\varkappa$ (more precisely, the solution will be valid for small values of the product $\varkappa H_0^2$). In order to find this solution, we substitute

$$\Psi = \Psi_\infty + \varphi = 1 + \varphi \quad \text{for} \quad |\varphi| \ll 1. \tag{4.29}$$

Then, in the first approximation, up to terms of order φA and φ^2, the system (4.18) assumes the form

$$\frac{\mathrm{d}^2\varphi}{\mathrm{d}z^2} = \varkappa^2(2\varphi + A^2), \quad \frac{\mathrm{d}^2A}{\mathrm{d}z^2} = A. \tag{4.30}$$

This system may be integrated at once, and its solution may be used for finding the next approximation, and so on. The corresponding solution, with the conditions (4.24) and (4.26), up to and including terms in H_0^3, has the form

$$\Psi = 1 + \frac{\varkappa H_0^2}{\sqrt{2}(2 - \varkappa^2)}\left(\frac{\varkappa}{\sqrt{2}}\exp\left(-2z\right) - \exp\left(-\sqrt{2}\varkappa z\right)\right),$$

$$A = -H_0\exp(-z) - \frac{\varkappa H_0^3}{\sqrt{2}(2 - \varkappa^2)}$$

$$\times \left\{\frac{\varkappa}{4\sqrt{2}}\exp\left(-3z\right) - \frac{\exp\left(-(\sqrt{2}\varkappa + 1)z\right)}{\varkappa(\varkappa + \sqrt{2})}\right.$$

$$\left. - \frac{3\varkappa^3 + 3\sqrt{2}\varkappa^2 - 8\varkappa - 4\sqrt{2}}{4\sqrt{2}\varkappa(\varkappa + \sqrt{2})}\exp\left(-z\right)\right\}. \tag{4.31}$$

For $z = 0$, naturally, $\mathrm{d}\Psi/\mathrm{d}z = 0$, $H = H_0$ and

$$\Psi_0 = 1 - \frac{\varkappa H_0^2}{2(\varkappa + \sqrt{2})}, \quad A_0 = -H_0 - \frac{\varkappa(\varkappa + \sqrt{2})}{4(\varkappa + \sqrt{2})^2}H_0^3. \tag{4.32}$$

The biggest of the terms in Ψ neglected in (4.32) are of the order $\varkappa^2 H_0^4$, and in A of the order $\varkappa^2 H_0^5$. The field H_0 in the equilibrium state is less than or equal to the critical field H_{cb} for the superconductor, which in the new variables is $1/\sqrt{2}$ [see (4.17)]. According to (4.32), for $\varkappa = 0.165$ [see (4.23)] $\Psi_0 \geq 0.974$ (the equality applies when $H_0 = 1/\sqrt{2}$), and thus the application of (4.31) here is completely justified if it is sufficient to determine $(\Psi - 1)$ to a few percent. At the present time, such an accuracy in the measurement of δ_0 is far from having been reached.

Since from the experimental data it follows that $\varkappa \ll 1$, and also for a reason indicated below the solution of (4.18) is possible for another limiting case when $\varkappa \to \infty$ does not offer any intrinsic interest, we shall not discuss it.

If $\varkappa = 0$, then in the problem under discussion $\Psi \equiv 1$ for any H – this corresponds to the usual theory based on (4.1) with $\Lambda = \text{const}$. If $\varkappa > 0$, the solution exists only up to a certain 'second critical field' H_{c2}. The range of fields $H_{cb} = 1/\sqrt{2} < H < H_{cb2}$ represents a metastable (superheated) state in which the superconducting phase can exist, since it represents a relative minimum of the free energy, but the absolute minimum of free energy is already that corresponding to the normal phase. The more detailed investigation of this question, and a calculation of the dependence of the field H_{c2} on $\varkappa$, has not yet been carried through.

Let us now note that for $\varkappa \geq 1/\sqrt{2}$ a peculiar instability of the normal phase of the metal occurs. Indeed, suppose the whole metal is in equilibrium, and in the normal state, i.e., $H_0 = 1/\sqrt{2}$. Then it can be shown that for $\varkappa \geq 1/\sqrt{2}$ an instability appears, with respect to the formation of thin layers of the superconducting phase, in the sense that solution of (4.18) appears with $\Psi \neq 0$. In the fact, assuming that $\Psi \ll 1$, we can take $H = H_0 = \text{const}$ and the first equation (4.18) assumes the form

$$\frac{\mathrm{d}^2 \Psi}{\mathrm{d}z^2} = -\varkappa^2(1 - H_0^2 Z^2)\Psi. \tag{4.33}$$

This equation in its form coincides with Schrodinger's equation for the harmonic oscillator and is well known to have solutions for Ψ which vanish for $z = \pm\infty$ if $\varkappa = 2H_0(n + 1/2)$, where $n = 0, 1, 2, \ldots$

Since for the normal phase $H_0 \geq 1/\sqrt{2}$ the minimum value of $\varkappa$ for which solutions can appear is $1/\sqrt{2}$. The point $z = 0$ chosen in (4.33) is quite arbitrary, i.e., a 'parasitic' solution can appear anywhere, and indeed there occurs a certain instability of the normal phase connected with the fact that when $\varkappa > 1/\sqrt{2}$ the surface energy $\sigma_{ns} < 0$ (see the end of Sect. ??).

It has not been necessary to investigate the nature of the state which occurs when $\varkappa > \varkappa_0$, since from the experimental data, it is true somewhat preliminary and worked out on the basis of (4.22), it follows that $\varkappa \ll 1$. Leaving on one side the question of the true value of $\varkappa$, we must, in any case, because of the indicated instability of the solution, note that all results obtained by us are valid only for the case

$$\varkappa < \varkappa_0 = \frac{1}{\sqrt{2}}. \tag{4.34}$$

We may use (4.31) to investigate the dependence on the field strength of the penetration depth of a magnetic field in a bulk superconductor [9, 10]. In agreement with the experimental method of measurement [11, 12] we define the penetration depth of a magnetic field in a bulk superconductor in the following way:

$$\delta = \frac{1}{H_0} \int\limits_0^\infty H \, \mathrm{d}z = \frac{\delta_0}{H_0} \int\limits_0^\infty H \, \mathrm{d}z = \delta_0 \frac{|A_0|}{H_0}, \qquad (4.35)$$

where H_0 is the external field (field at $z = 0$) and in the first expression we use the usual and in the second and third the reduced units for H, H_0, A_0, and z. Substituting the field (4.31) into (4.35) we have (in the usual units)

$$\delta = \delta_0 \left\{ 1 + \frac{\varkappa(\varkappa + 2\sqrt{2})}{8(\varkappa + \sqrt{2})^2} \left(\frac{H_0}{H_{cb}} \right)^2 \right\} \equiv \delta_0 \left\{ 1 + f(\varkappa) \left(\frac{H_0}{H_{cb}} \right)^2 \right\},$$

$$\frac{\mathrm{d}\delta}{\mathrm{d}T} = \frac{\mathrm{d}\delta_0}{\mathrm{d}T} + f(\varkappa) \left(\frac{H_0}{H_{cb}} \right)^2 \left\{ \frac{\mathrm{d}\delta_0}{\mathrm{d}T} - \frac{2(\mathrm{d}H_{cb}/\mathrm{d}T)}{H_{cb}} \delta_0 \right\}. \qquad (4.36)$$

From this it is clear that the quantity δ_0, as already mentioned, represents the penetration depth in a weak field. The function $f(\varkappa)'$ grows monotonically with $\varkappa$ in such a way that $f(0) = 0$, $f(\infty) = 1/8$, and for $\varkappa \ll 1$ $f(\varkappa) \sim \varkappa/4\sqrt{2}$. Thus, for $H_0 = H_{cb}$, even for $\varkappa = 1/\sqrt{2}$, $\delta = 1.07 \, \delta_0$, and for $\varkappa = 0.165$, $\delta = 1.028 \, \delta_0$. If, as was the case in (4.12), measurements of δ are carried out using a weak and slowly varying field H_1 in the presence of a strong field H_0, then $(\delta - \delta_0)/\delta_0 = 3f(\varkappa)(H_0/H_{cb})^2$, i.e., the effect is tripled. We see that the expected change of δ with H for mercury, for which according to our estimate $\varkappa = 0.165$, is very small and lies outside the limits of the accuracy of measurements achieved in [12] (the data of [11] as regards the dependence of δ on H are probably for reasons indicated in [12] not true; this is also evident from the fact that in [11] for a number of cases δ varies as H_0 rather than as H_0^2 (even in weak fields, where it must necessarily vary as H_0^2) since it is an even function of H_0. As we shall see in Sect. ?? for thin superconductors, the dependence of δ on H_0 is much bigger than for bulk ones and may be observed in experiments of the type described in [10] (thus it is possible that the dependence of δ on H in [10] is real, which does not contradict the absence of a noticeable effect in [12]).

4.4 The Surface Energy at the Boundary of the Superconducting and Normal Phases

For the calculation of σ_{ns} we must find the solution of (4.18) for a superconducting half-space limited by a half-space consisting of the normal phase of the same metal. Since the only difference between two phases is that in the one $\Psi \neq 0$ and in the other $\Psi = 0$, it is reasonable to suppose that the transition between the two phases takes place continuously in some transition layer. It can be shown that our equations have just such a continuous smooth solution, and do not, for instance, lead to a solution satisfying the conditions of the problem in which the function Ψ vanishes suddenly at some point. Thus, the transition from the superconducting phase to the normal phase takes place in a transition layer, in which for $z = \infty$ we have the superconducting phase, and for $z = -\infty$ the normal phase. This means that we must seek a solution of (4.18) with the boundary conditions

$$\Psi = \Psi_\infty = 1, \quad H = A = \frac{d\Psi}{dz} = 0 \quad \text{when} \quad z = \infty; \tag{4.37}$$

$$\Psi = \frac{d\Psi}{dz} = 0, \quad H = H_0 = \frac{1}{\sqrt{2}}, \quad A = H_0 z + \text{const} \quad \text{when} \quad z = -\infty.$$

In fact, of course, the transition layer has a breadth of the order of δ_0 (more precisely, as we shall see below, of the order of $\delta_0/\varkappa$), just as the magnetic field in a superconductor falls to zero in a distance of the order of δ_0, although, strictly speaking, it vanishes only at $z = \infty$.

Substituting (4.28) into (4.19), we obtain an expression for the surface energy σ_{ns},

$$\sigma_{ns} = \frac{H_{cb}^2}{2\pi} \delta_0 \int_{-\infty}^{\infty} \left\{ \frac{1}{2} - (1 - A^2)\Psi^2 + \frac{1}{2}\Psi^2 - H_0 H \right\} dz$$

$$= \frac{H_{cb}^2}{2\pi} \delta_0 \int_{-\infty}^{\infty} \left\{ \frac{1}{\varkappa^2} \left(\frac{d\Psi}{dz} \right)^2 + H^2 - H_0 H \right\} dz, \tag{4.38}$$

where (4.28) has been used, H_0 has been made equal to $1/\sqrt{2}$, and all quantities under the integral sign are expressed in reduced units.

In view of the fact that the general case involves the solution of (4.18) we can give an analytical expression for σ_{ns} only for a sufficiently small $\varkappa$. In this case, in the superconducting phase for a large z (far from the transition region) $\Psi = 1 - \text{const} \exp(-\sqrt{2}\varkappa z)$ [see (4.30) and (4.31)], i.e., changes only slowly with z. Consequently, we shall seek a solution of the second equation (4.18) in the form

$$A = \exp\left\{ -\int \Psi \, dz \right\}. \tag{4.39}$$

It is easy to see that this solution is valid for

$$\left| \frac{d}{dz} \left(\frac{1}{\Psi} \right) \right| \ll 1. \tag{4.40}$$

Substituting (4.39) into (4.28) we find that $d\Psi/dz = \varkappa(1 - \Psi^2)/\sqrt{2}$; i.e.,

$$\Psi = \tanh\left(\frac{\varkappa z}{\sqrt{2}} \right), \tag{4.41}$$

provided that the origin of z is suitably chosen. It should be noted that (4.41) is at the same time the strict solution of (4.18) for $\Psi(z)$ in the absence of an external field, and subject to the condition that $\Psi(\infty) = 1$ and $\Psi(0) = 0$. From (4.40) it is evident that (4.41) in the presence of a field applies as long as the inequality

$$\varkappa \ll \sqrt{2}\sinh^2\left(\frac{\varkappa z}{\sqrt{2}}\right) \tag{4.42}$$

is satisfied. With this condition, and taking into account (4.39) and (4.41), we find that

$$A = \exp\left\{-\int \Psi\,dz\right\} = C\exp\left[-\frac{\sqrt{2}}{\varkappa}\ln\cosh\left(\frac{\varkappa z}{\sqrt{2}}\right)\right],$$

$$H = \frac{dA}{dz} = -\Psi A = -A\tanh\left(\frac{\varkappa z}{\sqrt{2}}\right). \tag{4.43}$$

For $\varkappa z \ll 1$,

$$\Psi = \frac{\varkappa z}{\sqrt{2}}, \quad A = C\exp\left[-\frac{\varkappa z^2}{2\sqrt{2}}\right], \quad \varkappa z^2 \gg 1, \tag{4.44}$$

where the inequality $\varkappa z^2 \gg 1$ is obtained from (4.42). It is evident that the approximation (4.44) is valid if $1/\varkappa \gg 1/\sqrt{\varkappa}$; these inequalities may be satisfied if $\varkappa$ is sufficiently small. For an estimate of the constant C in (4.43) and (4.44), we take account of the fact that $H \leq H_0 = 1/\sqrt{2}$, and consequently $|A| \leq 1/\sqrt{2}\tanh(\varkappa z/\sqrt{2})$, or, if $\varkappa z \ll 1$, $|A| \sim 1/\varkappa z$ in the region where $H \sim 1/\sqrt{2}$. From this, taking into account that (4.44) still applies as regards an order of magnitude for $z \sim \varkappa^{-1/2}$, we find that $C \sim \varkappa^{-1/2}$. Thus, at the boundary of the region of validity of (4.41) to (4.43) $A \gg 1$ (since $\varkappa \ll 1$). But, if $A \gg 1$, (4.18) simplify and assume the form

$$\frac{d^2\Psi}{dz^2} = \varkappa^2 A^2 \Psi; \quad \frac{d^2 A}{dz^2} = \Psi^2 A. \tag{4.45}$$

Introducing the variables $\zeta = z\sqrt{\varkappa}$, $\varphi = \Psi/\sqrt{\varkappa}$ and $B = A\sqrt{\varkappa}$, we obtain from (4.45) the universal equations

$$\frac{d^2\varphi}{d\zeta^2} = \varphi B^2; \quad \frac{d^2 B}{d\zeta^2} = \varphi^2 B, \tag{4.46}$$

which, likewise, cannot be integrated analytically, but must be solved numerically once and for all. However, there is no need even to do this since it is easy to see that the contribution to σ_{ns} from the region where (4.45) and (4.46) are valid, i.e., the region $-\infty < z \lesssim \varkappa^{-1/2}$ is of the order $\varkappa^{-1/3}$. Similarly, the contribution from the region $\varkappa^{-1/2} < z < \infty$ is of the order of $\varkappa^{-1}$. Indeed, substituting into (4.38) both (4.41) and (4.43), it is easy to see that the contribution to σ_{ns} from the terms $H^2 - H_0 H$ under the integral sign is a quantity of order $\varkappa^{-1/2}$, while

$$\int\limits_{1/\sqrt{\varkappa}}^{\infty} \frac{1}{\varkappa^2} \left(\frac{\mathrm{d}\Psi}{\mathrm{d}z}\right)^2 \mathrm{d}z = \frac{\sqrt{2}}{3\varkappa} + \text{terms of the order } \varkappa^{-1/2},$$

connected with the lower limit of the integral. In this way, apart from the terms of the order of $\varkappa^{-1/2}$, we have

$$\sigma_{ns} = \frac{\delta_0 H_{cb}^2}{3\sqrt{2}\pi\varkappa}, \quad \Delta = \frac{\sigma_{ns}}{H_{cb}^2/8\pi} = \frac{1.89\delta_0}{\varkappa}, \quad \text{if} \quad \sqrt{\varkappa} \ll 1. \tag{4.47}$$

It is especially important to emphasise that, for small values of $\varkappa$, $\sigma_{ns} > 0$, which is absolutely necessary, and the attainment of which was our main aim. For a sufficiently large $\varkappa$, on the other hand, $\sigma_{ns} < 0$ [this is apparent immediately from (4.38) since $H^2 < H_0 H$], which indicates that such large values of $\varkappa$ do not correspond to the usually observed state of affairs (as a result of a numerical integration it turns out that $\sigma_{ns} = 0$ when $\varkappa = 1/\sqrt{2}$). The value (4.23), assumed by us for mercury, is very small from all other points of view, but insufficiently small for the applicability of (4.47), since in this case $\sqrt{\varkappa} = 0.407$. The numerical integration for $\varkappa = 0.165$ leads to a value of about $6\delta_0$ for Δ, while, according to (4.47), $\Delta = 11.4\,\delta_0$.

The thickness of the transition layer is evidently of the order of $\delta_0/\varkappa$, i.e., about $10\,\delta_0$.

4.5 Superconducting Plates (Films)

The solution is one-dimensional for plane plates and films, as well as for a half space. Here it is of interest to calculate the critical magnetic field, H_c, for the destruction of superconductivity in films and the magnetic moment of the film in an arbitrary field H_0; moreover, when there is a total current J flowing through the film we have to find the critical value of the current J_c to destroy superconductivity, and also the dependence of J_c on a superimposed field H_0.

The critical field H_c, as is shown by thermodynamic considerations [2,3] is determined by the relations

$$\frac{H_c^2}{8\pi} = \frac{H_{cb}^2}{8\pi} - \frac{\sigma}{d}, \quad \sigma = \int\limits_0^d \left(\frac{H^2(z)}{8\pi} - \frac{H_c H(z)}{4\pi} + \Delta F\right) \mathrm{d}z, \tag{4.48}$$

in which the thickness of the plate is $2d$, the z-axis is perpendicular to the plate with $z = 0$ at its centre, H_{cb} is the critical field for a bulk superconductor, and ΔF is the contribution to the free energy density of the superconductor, resulting from penetration of the magnetic field. In the theory based on (4.1), $\Delta = \Lambda j_s^2/2$, and the substitution of (4.4) into (4.48) leads to (4.5), if surface energy is ignored. In our case, we obtain from (4.9) and (4.7)

$$\Delta F = F_{n0} + \alpha \Psi^2 + \frac{\beta \Psi^4}{2} + \frac{\hbar^2}{2m}\left(\frac{d\Psi}{dz}\right)^2 + \frac{e^2}{2mc^2}A^2\Psi^2 - \left(F_{n0} - \frac{\alpha^2}{2\beta}\right),$$

and thus

$$\sigma = \frac{H_{cb}^2 \delta_0}{4\pi} \times \int\limits_0^d \left\{\frac{1}{2} - (1 - A^2)\Psi^2 + \frac{1}{2}\Psi^4 + \frac{1}{\varkappa^2}\left(\frac{d\Psi}{dz}\right)^2 + H^2 - 2H_cH\right\}dz,$$

$$(4.49)$$

in which the new units are used in the integrand. The quantity (4.49) is denoted by σ since it is clear from (4.15) that it is equivalent to the surface energy integrated with the proper limits.

The magnetic moment of the film per unit area in an external field H_0, parallel to the film, is given by

$$\mu = \int\limits_{-d}^d \frac{H(z) - H_0}{4\pi}\,dz = \frac{1}{2\pi}\left(A(d) - H_0(d)\right) \qquad (4.50)$$

where, in the transition to the second expression, it has been taken into account that for a film without a total current in an external field $H(z) = H(-z)$. $A(d)$ has been substituted for $\int_0^d H(z)\,dz$, since the potential A will be chosen below in such a way that $A(0) = 0$. Equation (4.50) can be obtained, either from the fact that the work of magnetisation of the film is given by

$$-\mu H_0 = \frac{H_0}{4\pi}\int\left(H_0 - H(z)\right)dz$$

or directly, from the fact that the field $H(z)$ plays the role of magnetic induction $B(z)$, and thus the expression $(H(z) - H_0)/4\pi$ is equivalent to the magnetisation $M = (B - H)/4\pi$.

For the determination of H_c, μ and J_c we must find the solution of (4.18) with the boundary conditions

$$\frac{d\Psi}{dz} = 0, \quad H = H_0 \pm H_J, \quad H_J = \frac{2\pi}{c}J, \quad \text{when} \quad z = \pm d. \qquad (4.51)$$

Here, H_0 is the external magnetic field directed along the y-axis, J is the total current $(J = \int_0^d j\,dz$, where j is the current density) flowing along the film in the direction of the negative x-axis, and $2H_J = 4\pi J/c$ is the difference between the values of the total field on both sides of the film, due to the current J. If the current J and the field H_0 are not mutually perpendicular, then there are two non-vanishing components of the potential A (in fact, A_x and A_y) instead of the single component A_x in the case considered above. We should then have, instead of (4.18), a system of initial equations of the form

$$\frac{d^2\Psi}{dz^2} = \varkappa^2\left(-(1 - A_x^2 - A_y^2)\Psi + \Psi^3\right), \quad \frac{d^2A_x}{dz^2} = \Psi^2 A_x, \quad \frac{d^2A_y}{dz^2} = \Psi^2 A_y.$$

$$(4.52)$$

These equations have to be solved for the conditions

$$H_x = H_{x0}, \quad H_y = H_{y0} \pm H_J, \quad H_J = \frac{2\pi J}{c}, \quad \frac{d\Psi}{dz} = 0, \quad \text{when} \quad z = \pm d.$$
(4.53)

The axes have now been chosen in such a way that the total current has a component only along the x-axis and, consequently, the field H_J is directed along the y-axis; H_{x0} and H_{y0} are the components of the external field along the x and y-axes.

For sufficiently thick plates, i.e., when $d \gg \delta_0$, the value H_c may be immediately obtained from the results of Sect. ?? by allowing d to tend to infinity in (4.48). Thus, substituting (4.31) into (4.48) we have, for $d \gg \delta_0$

$$\frac{H_c}{H_{cb}} = 1 + \frac{\delta_0}{2d}\left(1 + \frac{1}{2}f(\varkappa)\right).$$
(4.54)

Here, $f(\varkappa) = \varkappa(\varkappa + 2\sqrt{2})/8(\varkappa + \sqrt{2})^2$, the same function as in (4.36); (4.54) is valid up to terms of the order $(\delta_0/d)^2$. To the same approximation in the usual theory [1, 2], we should obtain (4.54) with $\varkappa = 0$ [see (4.5)]. Taking (4.36) into account, (4.54) may be written in the form

$$\frac{H_c}{H_{cb}} = 1 + \frac{\delta_0}{2d} + \frac{\Delta\delta}{4d},$$
(4.55)

where $\Delta\delta = \delta(H_{cb}) - \delta_0$.

For films of arbitrary thickness the solution of (4.18) must be carried out again. The solution of (4.31) suggests that, for thin plates as well as for thick ones, the function Ψ changes only slowly with z, if $\varkappa$ is small. Starting from this supposition, which is subsequently justified, we suppose that

$$\Psi = \Psi_0 + \varphi, \quad |\varphi| \ll \Psi_0, \quad \text{and} \quad \varphi = 0 \quad \text{when} \quad z = 0.$$
(4.56)

Then (4.18), in the first approximation, assumes the form

$$\frac{d^2\varphi}{dz^2} = \varkappa^2\{\Psi_0^3 - \Psi_0 + (3\Psi_0^2 - 1)\varphi + A^2\Psi_0\}, \quad \frac{d^2A}{dz^2} = \Psi_0^2 A.$$
(4.57)

From the second of the equations in (4.57), taking account of the boundary conditions (4.51), we find the values of A and H to be

$$A = \frac{H_0 \sinh \Psi_0 z}{\Psi_0 \cosh \Psi_0 d} + \frac{H_J \cosh \Psi_0 z}{\Psi_0 \sinh \Psi_0 d},$$

$$H = \frac{dA}{dz} = \frac{H_0 \cosh \Psi_0 z}{\cosh \Psi_0 d} + \frac{H_J \sinh \Psi_0 z}{\sinh \Psi_0 d}.$$
(4.58)

Substituting (4.58) in the first of the equations in (4.57), we find φ, and from the requirement that for $z = \pm d$, $d\varphi/dz = 0$, we obtain a transcendental equation determining Ψ_0.

As we shall see in practice we may with sufficient accuracy put $\varkappa = 0$. We shall, therefore, give the expression for φ and the equations for Ψ_0 for the case $\varkappa \neq 0$, only when $H_J = 0$, i.e., for a film in an external field. In this special case

$$\varphi = -\frac{\Psi_0(\Psi_0^2 - 1)}{3\Psi_0^2 - 1}\left\{1 - \cosh \varkappa z \sqrt{3\Psi_0^2 - 1}\right\}$$

$$+ \frac{\varkappa H_0^2}{2\Psi_0^2 \sqrt{3\Psi_0^2 - 1}\cosh^2 \Psi_0 d}\left\{\frac{1 - \cosh \varkappa z \sqrt{3\Psi_0^2 - 1}}{\varkappa \sqrt{3\Psi_0^2 - 1}}\right.$$

$$\left. - \frac{\varkappa\sqrt{3\Psi_0^2 - 1}(\cosh \varkappa\sqrt{3\Psi_0^2 - 1}\,z - \cosh 2\Psi_0 z)}{4\Psi_0^2 - \varkappa^2(3\Psi_0^2 - 1)}\right\}, \tag{4.59}$$

$$\Psi_0^2 - 1 = 2H_0^2\left\{1 - \frac{\sinh 2\Psi_0 d}{2\Psi_0 d}\frac{\varkappa d\sqrt{3\Psi_0^2 - 1}}{\sinh \varkappa d\sqrt{3\Psi_0^2 - 1}}\right\}$$

$$\times\left\{\cosh^2 \Psi_0 d[4\Psi_0^2 - \varkappa^2(3\Psi_0^2 - 1)]\right\}^{-1}. \tag{4.60}$$

In the limiting case $\varkappa = 0$, for arbitrary H_0 and H_J, naturally $\varphi = 0$ and

$$\Psi_0^2(\Psi_0^2 - 1) = \frac{H_0^2\left[1 - (\sinh 2\Psi_0 d)/(2\Psi_0 d)\right]}{2\cosh^2 \Psi_0 d}$$

$$- \frac{H_J^2\left[1 + (\sinh 2\Psi_0 d)/(2\Psi_0 d)\right]}{2\sinh^2 \Psi_0 d}. \tag{4.61}$$

Let us note that for $\varkappa = 0$ the equation for $\Psi = \Psi_0 = $ const may be immediately obtained from the condition of minimum free energy, i.e., from the condition $d\sigma/d\Psi = 0$. It is clear from (4.48), that this condition gives

$$\Psi_0^2 - 1 = -\frac{1}{d}\int_0^d A^2\,dz,$$

which leads to (4.61).

Let us now discuss, in somewhat more detail, the destruction of superconductivity in a film by an external field in the absence of a total current. If $\varkappa = 0$, then $\Psi = \Psi_0 = $ const and (4.58) applies with $H_J = 0$. Substituting this solution in (4.48) we easily find (in the usual units)

$$\left(\frac{H_c}{H_{cb}}\right)^2 = \frac{\Psi_0^2(2 - \Psi_0^2)}{1 - \tanh \eta/\eta}, \quad \text{where} \quad \eta = \frac{\Psi_0 d}{\delta_0}. \tag{4.62}$$

In this case ($\varkappa = 0$, $H_J = 0$), (4.61) becomes, when $H_0 = H_c$, or in the usual units when $H_0 = H_c/\sqrt{2}H_{cb}$

$$\left(\frac{H_c}{H_{cb}}\right)^2 = \frac{4\Psi_0^2(\Psi_0^2-1)\cosh^2\eta}{1-(\sinh 2\eta)/2\eta}, \quad \text{where} \quad \eta = \frac{\Psi_0 d}{\delta_0}. \tag{4.63}$$

From (4.62) and (4.63), from the measured values of H_c/H_{cb}, and from d, we can determine Ψ_0 and δ_0. It is easy to see that for small values of η and for $H = H_c$, $\Psi_0 = 0$ and

$$\frac{H_c}{H_{cb}} = \sqrt{6}\,\frac{\delta_0}{d}. \tag{4.64}$$

Thus, in this case, we have a phase change of the second kind; with growth of the field Ψ_0 decreases and at the transition point $\Psi_0 = 0$. As is evident from (4.61), for $H_J = 0$, up to terms of order d^2 (taking into account that $H_0^2 d^2$ may be of the order unity) we have

$$\Psi_0^2 = \frac{1-(H_0/H_{cb})^2(d^2/6\delta_0^2)}{1-(2/15)(H_0/H_{cb})^2(d^4/\delta_0^4)}.$$

The transition to the normal state is one of the second kind for $d \le d_c$, where it is easily shown from (4.62) and (4.63) that

$$d_c = \frac{\sqrt{5}\,\delta_0}{2}. \tag{4.65}$$

The point $d = d_c$ is a kind of critical Curie point [7], and for $d > d_c$ we have a transition of the first kind; i.e., for $H_0 = H_c$, $\Psi_0 > 0$ and there is a latent heat of transition (for $d < d_c$ and for $H_0 = H_c$ we have a jump in the specific heat; the specific heat of thin plates evidently depends on H_0).

The penetration depth of the field is clearly from (4.58) the quantity

$$\delta = \frac{\delta_0}{\Psi_0}, \tag{4.66}$$

and we see that for sufficiently thin specimens the penetration depth may be appreciably larger than for the bulk metal when $H_0 \sim H_c$. Here [see (4.50) and (4.58) with $H_J = 0$]

$$\mu = -\frac{H_0 d}{2\pi}\left(1 - \frac{\delta}{d}\tanh\frac{d}{\delta}\right) = -\left(\frac{H_0 d}{2\pi}\right)\left\{\frac{1}{3}\left(\frac{d}{\delta}\right)^2 - \frac{2}{15}\left(\frac{d}{\delta}\right)^4 + \dots\right\}. \tag{4.67}$$

From measurements of μ we may find the penetration depth δ, which, according to (4.66) and (4.61), depends on H_0.

For $\varkappa \ne 0$ all the expressions become exceedingly complicated in the general case. However, for small values of $\varkappa$, which are the only ones which interest us, and for not too large values of d, we may expand all the expressions as a series in $\varkappa d$. The result is, that in the range of thicknesses for which the transition is of the second kind, (4.64) must be replaced by the expression

$$\left(\frac{H_c}{H_{cb}}\right)^2 = 6\left(\frac{\delta_0}{d}\right)^2 - \frac{7}{10}\varkappa^2 + \frac{11}{1400}\varkappa^4\left(\frac{d}{\delta_0}\right)^2 \dots \tag{4.68}$$

The value of d_c is then given by

$$d_c^2 = \frac{5}{4}\left(1 - \frac{7}{24}\varkappa^2 + \ldots\right)\delta_0^2. \tag{4.69}$$

If we take for $\varkappa^2$ the value (4.23) then in practice it is hardly necessary to take into account the term in $\varkappa^2$ in (4.68), (4.69) and in the analogous expressions.

The only experimental data on the destruction of superconductivity in films by an external field suitable for a quantitative discussion are those given in [4], and refer to mercury. The scatter of points, however, even in these measurements was rather large, and moreover, in the absence of tables, the values of H_c/H_{cb} had to be taken from graphs; nevertheless, the chief source of error is due to the fact the thickness of the films indicated in [4] is some sort of average value and may, especially for the thin films, differ considerably from the thickness d entering in our formulae, in which it is assumed, of course, that the film is ideally uniform.

In Table 4.1 we reproduce the values of δ_0 obtained with the aid of (4.64) on the basis of the data for H_c/H_{cb} as a function of d given in [4]; the values shown in brackets are those for which the calculation from (4.64) is already invalid since $d > d_c$. Underneath these values in brackets are put the values of δ_0 obtained directly from (4.62) and (4.63).

In the last column are shown the values of $2d_c$ obtained from (4.65) with the help of the minimum values of δ_0 in the corresponding line. From Table 4.1, as also shown directly from the graph given in [4], showing the dependence of $\ln(H_c/H_{cb})$ on $\ln 2d$, it is clear that there is a sharp break in the course of this dependence, which sets in as d passes through d_c (in Table 4.1 the single values of δ_0 and the values in brackets according to (4.64) are simply quantities proportional to $(H_c/H_{cb})d$; this product falls as d rises to d_c and for $d > d_c$ the sharp rise begins). We are inclined to regard this behaviour as confirmation of the conclusion that the character of the transition is different for $d < d_c$ and $d > d_c$. The fall of the values of δ_0 with rise of d, clearly evident from Table 4.1 for $d > d_c$, may be completely explained by the already mentioned difference between the values of d indicated in [4] and the effective values d_{eff}. Alternatively, it is evident that the thinner the film the more will d_{eff} depart from d, and that $d_{\text{eff}} < d$. The observed dependence of δ_0 on d for $d < d_c$ is in agreement with this picture, but we can see no reason for the increase with d of the values of δ_0 calculated according to (4.62) and (4.63) when $d > d_c$.

We must, however, bear in mind two considerations. Firstly, the whole of our scheme based on the expansion of F_{s0} and (4.10) in powers Ψ^2 up to the terms in Ψ^4 is, generally speaking, valid only in the region close to T_c in which (4.8) and the equation

$$\delta_0^2 = \frac{\text{const}}{T_c - T} = \frac{\delta_{00}^2}{1 - T/T_c}, \tag{4.70}$$

are valid, where δ_{00} is a certain constant [see (4.20) and (4.7)]. For mercury the region where (4.8) is valid [and therefore (4.70) should be applicable] lies

Table 4.1. The values of δ_0 for mercury (δ_0 and d in units of 10^{-5}cm)

T, K	2d								$2d_s = \sqrt{5}\delta_{0\,\mathrm{min}}$
	0.596	0.840	1.178	1.423	1.690	2.400	4.390	10.880	
4.13	5.13	4.61	4.07	4.17	3.80	3.37	3.08	(3.72)	6.9
								3.56	
4.12	4.12	4.06	3.47	3.36	3.27	3.11	2.72	(3.52)	6.1
								3.30	
4.10	3.47	3.38	2.87	3.02	2.79	2.53	2.28	(3.21)	5.1
								2.50	
4.05	2.66	2.62	2.32	2.27	2.08	1.86	(1.80)	(2.86)	4.0
							1.80	1.95	
4.00	2.28	2.31	1.92	1.82	1.76	1.56	(1.57)	(2.72)	3.5
							1.57	1.70	
3.80	1.69	1.62	1.40	1.28	1.24	1.10	(1.31)	(2.63)	2.5
							1.15	1.39	
3.60	1.27	1.24	1.08	0.99	0.98	(0.87)	(1.23)	(2.50)	1.95
						0.87	0.99	1.16	
3.00	1.10	1.10	0.92	0.84	(0.83)	(0.77)	(1.16)	–	1.61
					0.83	0.72	0.84		
2.50	0.92	0.94	0.86	0.80	(0.75)	(0.73)	(1.13)	(2.45)	1.48
					0.75	0.66	0.78	1.0	

between T_c and $T \sim 3,80$–4.0 K. For smaller values of T we must, in general, take into account higher terms in the series expansion of F_{s0} [i.e., terms in Ψ^5 etc. in (4.18)] and the application of all the formulae obtained without the substitution of $|\alpha|/\beta$ by $(d\alpha/dT)_c(T_c - T)/\beta_c$ is possible only if the nonlinear dependence of $|\alpha|/\beta$ on $(T_c - T)$ is more important than the influence of terms in Ψ^6, etc. Such a situation is possible, but it could not be assumed to occur unless it were demonstrated by an analysis of sufficiently extensive experimental data; this is not possible at present, owing to the absence of the latter. In view of what has been said, the data of Table 4.1 for $T < 3.80$ K may be distorted.

Secondly, we must bear in mind is that T_c varies considerably from film to film; in [4] all the data was reduced to $T_c = 4.167$ K and this operation, evidently inaccurate for $T = 4.12$ K and $T = 4.13$ K, may also influence the data in Table 4.1 at lower temperatures. The whole question clearly requires a more detailed experimental investigation; for the moment, we shall take for δ_0 the lowest of the values in Table 4.1 and compare them with the data obtained by other methods [10, 12]. In doing this we must consider the fact that in [12] the quantity directly measured was only $\delta_0 - \delta_0(2.5\,\mathrm{K})$, and that δ_0 was calculated by means of an extrapolation which does not appear a priori

Table 4.2.

T, K	δ_0, from Table 4.1	$\delta_0 - \delta_0(2.5\,\mathrm{K})$, from Table 4.1	δ_0, from [10]	$\delta_0 - \delta_0(2.5\,\mathrm{K})$, from [10]	δ_0, from [12]	$\delta_0 - \delta_0(2.5\,\mathrm{K})$, from [12]
4.13	3.08	2.42	4.08	3.28	2.28	1.82
4.12	2.72	2.06	3.57	2.77	2.04	1.58
4.10	2.28	1.62	2.80	2.00	1.72	1.26
4.05	1.80	1.14	2.34	1.54	1.31	0.85
4.00	1.56	0.90	1.95	1.15	1.10	0.64
3.80	1.10	0.44	1.38	0.58	0.77	0.31
3.50	0.87	0.21	–	–	0.61	0.15
3.00	0.72	0.06	–	–	0.50	0.04
2.50	0.66	0.00	0.80	0.00	0.46	0.00

valid. The values of δ_0 obtained in [10] are based on the previous measurements with the colloids, and are likewise inaccurate; here also the measured quantity was $\delta_0 - \delta(2.5\,\mathrm{K})$. As can be seen from Table 4.2, in which all the quantities must be multiplied by 10^{-5}cm within the limits of the accuracy achieved up to the present time, the data of Table 4.1 coincides with those obtained by other methods (we must especially emphasise that the data of [12] relates to bulk specimens).

Assuming for δ_0 the values indicated in the second column of Table 4.1, we may calculate $\varkappa$ with the help of (4.22), taking also into account the fact that for mercury close to T_c, $H_{cb} = 187(T_c - T)$. Thus, if we use the most reliable value of δ_0 at $4\,\mathrm{K}$ we obtain (4.23). Using the value of δ_0 indicated in [12] for mercury and for tin, we obtain $\varkappa \sim 0.15$.

Let us now turn to the question of the destruction of the superconductivity of a film by a current. For $\varkappa = 0$ the function Ψ_0 in the presence of a current is given by (4.61), which for $d \ll 1$ takes the form

$$\Psi_0^2 = 1 - \frac{H_0^2 d^2}{3} - \frac{H_J^2}{\Psi_0^4 d^2}. \tag{4.71}$$

The field H_J as a function of Ψ_0 becomes zero for $\Psi_0 = 0$ and for some nonvanishing value of Ψ_0 (if $H_0 = 0$ then $H_J = 0$ for $\Psi_0 = 1$); between these two values of Ψ_0, H_J exhibits a maximum. In other words, the function Ψ_0 for the given H_J may, according to (4.71), have two values. It is easy to see that the superconductivity of a film is stable only so long as the field H_J of the current grows with the decrease of Ψ_0 (in this case, the free energy is less than that corresponding to the same H_J, but a lower value of Ψ_0). The critical field H_{Jc} is determined from the condition $dH_J/d\Psi_0 = 0$, which leads to the relation

$$\frac{H_c}{H_{cb}} = \frac{2\sqrt{2}}{3\sqrt{3}} \frac{d}{\delta_0} \left[1 - \left(\frac{H_0}{H_c^2} \right)^2 \right]^{3/2}, \tag{4.72}$$

where H_c is the critical field of a given film in the absence of a current, H_0 is the external field and J_c is the critical current $(H_{Jc} = (2\pi/c)J_c)$. In the absence of a field H_0 we have

$$\frac{H_{Jc}}{H_{cb}} = \frac{2\sqrt{2}}{3\sqrt{3}} \frac{d}{\delta_0}. \tag{4.73}$$

For the case of arbitrary relative orientations of H_0 and H_J, where we must use (4.52) with the boundary conditions of (4.53), it is easy to see that we obtain the previous equations (4.61), (4.71), and (4.72) with $H_0^2 = H_{x0}^2 + H_{y0}^2$ (the current J is directed along the negative x-axis, the field H_J along the y-axis). It should be noted that it follows from (4.64) and (4.73), for sufficiently thin films, that

$$H_c H_{Jc} = \frac{4}{3} H_{cb}^2. \tag{4.74}$$

Thus, although the values of H_c and H_{Jc} for thin films may be greatly different from H_{cb}, the product $H_c H_{Jc}$, which is equal for a massive specimen to H_{cb}^2 is multiplied by a factor $4/3$ for the very thinnest films. Relations (4.72) and (4.73) are in qualitative agreement with experiments from which, however, it is impossible to draw quantitative conclusions.

Summarising, we may indicate that for an experimental verification of the theory, there is a whole number of possibilities: the measurement of the critical field and current for films [(4.62)–(4.64) and (4.73)]; the measurement of the influence of field on the critical current [see (4.72)]; the measurement of the magnetic moment [see (4.66) and (4.67)]; the measurement of σ_{ns} and, finally, the measurement of $\delta(H_0)$ for bulk superconductors [see (4.36) and (4.54)]. However, in working with films, a direct determination of $\varkappa$ (if it is really small) is, in practice, apparently not possible. Thus, for the determination of $\varkappa$ not using (4.23) we must either determine σ_{ns} – a quantity which is particularly sensitive to $\varkappa$ – or carry out exact (~ 1 percent) measurements with bulk superconductors of the influence on δ of fields of the order of H_{cb}.

4.6 Comments on the Present Edition

In 1950, the publication of articles by Soviet authors in English was impossible. As far as I remember, neither did we send Russian offprints to anybody abroad. However, the article was noticed by D. Shoenberg (Cambridge, UK); he translated it into English and, probably, showed his translation to a number of foreign physicists. The article was published in English only in the collection of the works of L.D. Landau [Landau L.D. *Collected Papers* (Oxford: Pergamon Press, 1965)]. The present edition completely reproduces the Russian text (*ZhETF* **20**, 1064 (1950)), with only a few misprints corrected. This article was widely used; I myself used it as a basis for a number of works mentioned in the article 2 'Superconductivity and Superfluidity' published in

Fig. 4.1. Nobel Diploma of Nobel Prize Laureate in Physics 2003 V.L. Ginzburg

the present book. In the article written by L.D. Landau and myself few minor inaccuracies were noticed, but I do not remember where exactly (for instance, it seems that the formula which is not numbered and follows (4.64) is inexact). As far as I know, such errors are not numerous and are completely immaterial. It should also be taken into consideration that in this article of ours, written in 1950, and in some subsequent articles of mine, the effective charge e in the formula (4.9) and henceforward was taken to be equal to the charge of the electron e_0 rather than to the charge of the 'pair' $2e_0$. Moreover, critical fields H_{c1} and H_{c2} were marked, respectively, as H_{k2} and H_{k1}. We shall note that a certain amount of information on the Ginzburg–Landau theory is given in references cited in Chaps. 1 and 2 of the present edition. See also the following books: Shmidt, V.V. *The Physics of Superconductors: Introduction to Fundamentals and Applications* (Eds. P. Müller, A.V. Ustinov) (New York, Springer 1997), Tinkham, M. *Introduction to Superconductivity*, 2nd edn. (New York, McGraw Hill 1996), and Buckel, W., Kleiner, R. *Superconductivity* (Wiley–VCH, 2004).

According to the tradition of the Nobel Foundation, along with the Nobel Gold Medal prize-winners are also given the Nobel Diploma. The idea of the picture illustrating such a diploma and specially made by an artist is supposed to represent, in an artistic form, a symbol of the prize-winner's main achievement for which the Nobel Prize has been awarded. The idea chosen by the artist Niels G. Stenkvist, who illustrated the Nobel Diploma of V.L. Ginzburg, represented "the psi function" (Ψ is the order parameter) introduced by V.L. Ginzburg and L.D. Landau in the article ("On the Theory of Superconductivity" (Chap. 4 of this book).

References

1. F. London, *Une conception nouvelle de la supraconductibilite* (New conception of the superconductivity) Paris, 1937.
2. V.L. Ginzburg, *Superconductivity*, Publishing house of the Academy of Sciences of the USSR, 1946.
3. V.L. Ginzburg, Zhur. eksp. teor. fiz. **16**, 87, 1946.
4. E. Appleyard, J.R. Bristow, H. London and A.D. Misener, Proc. Roy. Soc. **172**, 540, 1939.
5. H. London, Proc. Roy. Soc. **152**, 650, 1935.
6. A.I. Shalnikov, J.E.T.P. **10**, 630, 1940.
7. L.D. Landau and E.M. Lifshitz, *Statistical Physics*, Clarendon Press, 1940.
8. V.L. Ginzburg, U.F.N. **38**, 490, 1949.
9. V.L. Ginzburg, J. Phys. **11**, 93, 1947.
10. M. Desirant and D. Shoenberg, Proc. Phys. Soc. **60**, 413, 1948.
11. A.I. Shalnikov and U.V. Sharvin, Zhur. eksp. teor. fiz. **18**, 102, 1948.
12. E. Laurmann and D. Shoenberg, Proc. Roy. Soc. **198**, 560, 1949.

5

'Nobel' Autobiography
(A Supplement to the Nobel Lecture, 2003)

Apart from several monographs on physics and astrophysics, I have published two books which are collections of various articles of a scientific, semi-scientific, and general, social and political nature[1] Some of these articles contain a great deal of biographical material, and so I do not have any wish to come back to my autobiography. However, it is the wish of the Nobel Foundation that the Nobel Prize winners, together with the lectures, write their autobiographies. I respect the wishes of the Foundation, so I am writing. Of course, I could confine myself to brief biographical information; writing a detailed autobiography is rather dangerous – it can bring reproach for 'exhibitionism' and immodesty. Nevertheless, I have decided to write in a rather detailed and frank way, as it corresponds to my habits and tastes. One more reason to justify this decision is that I am already 87 and will hardly ever have another occasion to write about myself and my views.

I was born in Moscow, on 4 October, 1916, that is, as far back as in Tsarist Russia (at that time even the calendar was different, so the date of my birth was then the 21st September). My father, Lazar' Efimovich Ginzburg, lived almost half of his life in the 19th century; he was born in 1863, that is, only two years after the abolition of serfdom in Russia. Having graduated from Riga Polytechnic, he was an engineer engaged in the purification of water and had a number of patents. My mother, Avgusta Veniaminovna Vil'dauer-Ginzburg, was a doctor; she was born in 1886, in Mitava (Latvia). I was the only child in the family. Mother died of typhoid in 1920; I only remember one episode at her bedside and at her funeral. Mother's younger sister, Rosa, who then started

---

[1] I mean the book "O Fizike I Astrofizike" (On Physics and Astrophysics) (Moscow: Byuro Kvantum, 1995); it is the third edition. The English translation: "The Physics of a Lifetime" (Springer, 2001). The second book is "O Nauke, o Sebe i o Drugikh" (About Science, Myself and Others) (Moscow: Fizmatlit, 2003): it is the third edition. [The English translation "About Science, Myself and Others" published at the publishing house Bristol: IOP Pubs. (2005).] These books are referred to here as (I) and (II), respectively.

Fig. 5.1. Parents of V.L. Ginzburg: Lazar Efimovich Ginzburg and Avgusta Veniaminovna Vil'dauer–Ginzburg

to live with us, did everything she could for me. She died in 1948. Father had
died still earlier (in 1942), when we were living in Kazan', where we had been
evacuated. Later I will tell about this period in a more detailed way. And here
I will only note that, as is well known, life in Russia during World War I, and
especially during the period of so called 'military communism,' and then later
in the USSR was hard; in many cases even very hard. Nevertheless, in Moscow,
where we lived all the time, with the exception of about two years (1941–1943)
spent in Kazan', material conditions were better than in the majority of other
regions of the country. However, one of my childhood memories is a wagon
loaded with half-covered coffins with dead bodies and pulled by a horse past
our house in the center of Moscow. Another memory is of buying fresh meat
'for the kid', which turned out to be the meat of a dog. Normally, dogs had
never been eaten in Central Russia. However, in general we did not starve,
but we lived in a so-called communal apartment – two more families had
been placed into my father's four-room apartment after the revolution. Still,
it is not these hardships, more or less general, but my loneliness, that really
sticks in my mind. It was exacerbated by my being sent to school only at
the age of 11 (in 1927, when I entered the 4th form). My parents (or, to
be more exact, father and aunt) must have been afraid that school in the
Soviet times had become quite bad, and sending children to school was not
obligatory. I am not acquainted with the general state of education in the
USSR in the 1920s, but school No. 57, belonging to the Sokol'niki district

Fig. 5.2. Vitaly Ginzburg, 3 years and 4 months

department of education, which I entered, did not seem bad to me. It was a former French grammar school, with many good, old teachers still working there. True, as far as I remember, history was practically not taught at all, being substituted with a rendering of the contents of some reports of comrade Stalin, and, it seems, some other material like that. What was really bad about school came later. In 1931, just at the time when I finished the 7th form, yet another school reform happened, and the full high school (I do not remember whether it had 9 or 10 forms) was abolished. After the 7th form pupils were to enter vocational schools (FZUs), which were supposed to train factory workers to be skilled. After a FZU one could, in principle, enter directly higher educational establishments (VUZs), including universities – through 'rabfaks', that is, worker preparatory departments. I am putting it somewhat vaguely, because, not at all willing to enter a FZU, I did not follow this way. Thus, for a while I, then a boy of 15, remained rather lost and unhappy. My aunt was working in an organization dealing with the purchases of foreign scientific literature. One of their customers was Evgeni Bakhmet'ev, a professor at a technical VUZ. He was a very picturesque figure – a former submarine sailor, a Bolshevik, who became a specialist in the field of X-ray structure analysis. His fate, like the fates of many other people of such a kind, was tragic – later he died, crushed by the millstones of Stalins's terror. However, in 1931 he helped me to get a job as a laboratory assistant in the X-ray laboratory of the technical VUZ where he was teaching. I will not go into details. I will confine myself to saying that the key figure in the laboratory was Veniamin (Venya) Tsukerman, quite a young man (only three years older than me) with a truly uncommon inventiveness and initiative. With him and his school friend and peer, Lev Al'tshuler, we communicated and worked together for about two years. Both Tsukerman and Al'tshuler became physicists; from 1946 on they were very close to their superior, Y.B. Khariton, who headed the main center where the atomic and hydrogen bombs were created (now it is the town of Sarov, as this place was called as far back as in the Tsarist time; after a nuclear center was created there it was called by some code name, which I have already forgotten, and later its name was Arzamas-16). Tsukerman and Al'tshuler have a lot of awards; many works of theirs are already declassified and well-known. Both of them are people with interesting fates, in part already described in literature. Giving a more detailed description of them here is evidently impossible (I will only say that Tsukerman died in 1993, and on 9th November, 2003 we celebrated Al'tshuler's 90th birthday).

What I acquired at the laboratory was not so much some concrete knowledge as the taste for work, for physics, and for inventiveness. I also remember reading, with interest, the book by O.D. Khvol'son, "Fizika Nashikh Dnei" (*The Physics of Our Days*) – a popular writing about the achievements of physics, a rare book for that time. Anyway, I decided to become a physicist, especially since I did not have any talents, but was at least interested in physics. I decided to enter the physical department of the Moscow State University (MGU). Just at that time, in 1933, it became possible to enter the MGU by open competition (previously, for several years, the entrance procedure was somehow different – for instance, through rabfaks or in some other way, but certainly not by competition). However, for entering the university it was necessary to have finished the complete high school course, while I had finished only 7 forms. Thus, I had to learn in three months (partly with a tutor) what had to be learned in the senior forms. I am going into these details because all my life I have wished I had gotten a normal school education. And I want to warn those who think that school wisdom is not worth spending many years on. Indeed, though having, as I am convinced, merely average abilities, I managed to do in three months the program of three school years. But at what cost? The cost was, first of all, a total lack of automatism in elementary mathematics and in even the orthography of the Russian language. Specifically, at school I would have solved, for instance, 100 problems in algebra, trigonometry, etc., while on my own I would solved only about 10. The idea why solve more if it is not demanded have lead to a lack of skills in the further learning of mathematics. The same with orthography. Illiteracy 'in the people' was at that time being combatted though, and, as far as I know, there were some good results in that field. At the higher level, however, things were much worse. It suffices to say that already in the second year at the university we had a check on our Russian (a dictation), and half of the students of the year, including me, got unsatisfactory marks. After that, an obligatory course of Russian was introduced for those who had failed this exam. I doubt that it gave any noticeable benefit – making up for what has been missed in childhood is very difficult. I also know no foreign languages, although later I had to master English, but only in the limits necessary for my doing physics. I hope that the information given is not superfluous, both in terms of understanding the life in the USSR in the years described and in terms of my biography.

I did not fail any of the entrance exams at the MGU, in 1933, but on the whole I did not make out brilliantly. As a result, I was not admitted, although admitted were some people with slightly lower marks but more advantageous personal particulars (I was neither a member of the Young Communist League nor a worker, and my parents also were not proletarians, and so on). Still, this result was neither discrimination nor a sign of anti-Semitism, which flourished after the war. Some of my comrades who were not admitted either decided to wait for a year, but I had already left work and had somehow become keen on studying. That is why I entered the external department, which proved possible. And here again I painfully felt my weak spot. In 1934 I managed to get transferred to the internal department, to the second year, that is; I caught up with my fellow students and started to study like everybody. But I also learned how much richer and brighter their life was, with all sorts of optional courses and so on. Furthermore, strangely, I never got acquainted with the courses of astronomy and chemistry, which I did not have to do in the external department and somehow was not obliged to do when being transferred to the internal department.

I want to finish the story of the first stage of my life by mentioning just one of many episodes which show that man's fate is merely a chain of chances and is like a pathetic little boat in the sea waves, which may capsize at any moment. What I mean here is the following: at the second year at the MGU all students were enrolled in groups of two types, military and civil. Only men were placed in military groups, where they were trained and qualified as military officers. Therefore, when being transferred to the internal department

I was sent to a military hospital in order to decide where I should be placed. I was not particulary athletic (height 180 cm, weight about 60 kilograms). A military doctor, who seemed an elderly man to me, poked his finger at my neck, uttered the diagnosis 'struma' and sent me to a civil group. Struma is some swelling of the thyroid gland, which I had not noticed up to now. I am telling about that because practically all my fellow students who got into military groups were killed in the war. To finish with this unpleasant topic, I shall note that there were three more similar episodes. I cannot say that I was bursting to go to the front, but I was not in the least trying to avoid it. Telling about all these incidents would be too long, so I shall limit myself to only one. In 1941, after the beginning of the war (for the USSR it began on 22nd June, 1941) the Lebedev Physical Institute of the USSR Academy of Sciences (FIAN), where I was working at the time (and where I have been working up till now), was evacuated to Kazan' together with the greater part of the Academy of Sciences. Furthermore, what was quite natural for the war time, at least in our country, the workers of the Institute were used for different projects. For example, once we were sent to unload barges with logs or firewood on the Volga. By the by, everybody took part in that, for instance, I.E. Tamm (also mentioned in my Nobel Lecture). My task was bringing logs from the barge to the bank. I was wearing the so-called 'koza' (which means 'a goat') – something like a rucksack with a step at the rear. Such 'kozas' were widely used in Russia at one time for carrying heavy weights, and they are effective indeed. Two men put a log on a 'koza' and the one who was wearing it carried the log to the bank. I am still surprised what heavy logs can be carried in such a way. Still, I must have overstrained myself, and on the next day I had a bleeding at the throat, though not profuse. Some small blood vessel must have burst. I was sent to hospital, where some petrified spots were found in my lungs. The bleeding stopped, and, like the struma, it did not show itself anymore. At the same time, voluntary enlistment into some airborne landing forces was announced and I, then a YCL member, enlisted. But I was not admitted, although I had not mentioned any illnesses and, if I remember correctly, even did not know that I was on the medical books. By the by, I can say that I was not officially exempted from military service – at least, in the first years of the war. However, there was some order not to mobilize untrained graduates of some professions or some specialists in general. I was never conscripted, but, expecting to be conscripted all the time, I was trying to write my thesis as soon as possible. In peacetime I would not have hurried to do it, but then I wanted, though it looks a bit silly, to be able to finish before the draft what had been already started. I defended this thesis in the spring of 1942 (it was the D.Sc. thesis, as I had defended the Ph.D. thesis in 1940).

The life in Kazan' was hard; the four of us (my father, my aunt, my wife and I, as my daughter with her grandmother had been evacuated to another place) lived in one room. We felt rather cold and rather hungry. But I worked a lot, as did all my fellow workers. I was exploring the propagation of radio waves in the atmosphere and something else which seemed useful for the defense. I also tackled some other problems, as scientific life was going on.

However, let me come back to the student days. Starting from the second year, I did well in my studies and, so to say, enjoyed studying. I liked theoretical physics, but I thought, and still think, that my ability for mathematics is average, at most; I had difficulties solving problems and making calculations, while it was considered, and rather justly, that a theoretical physicist should be good at mathematics. In short, because of all that, when at the end of the third year or at the beginning of the fourth year (now I do not remember when it was) I had to choose my specialization, I did not dare to take up theoretical physics and chose optics. The choice was not accidental. At the MGU physical department at that time there were rather many really good specialists, but there were some retrogrades as well. They were struggling with each other, and this struggle was politically colored. Just a beginner at that time, I was far from their controversy. I remember only one public debate between the partisans of the contemporary physics and the so-called 'mechanicists', who criticized the theory of relativity and so on. It took place in 1935, or maybe the first half of 1936. I do not remember the date and I did not try to find it out, as I know the fate of B. Gessen – a participant of this dispute, a physicist and a philosopher, who was the dean of the physical department at one time. He was also an old Bolshevik (as they called those who joined the party whether before October 1917 or shortly after) and shared the lot of the majority of his comrades – arrested on 21 August, 1936 and sentenced to be shot on 20 December, 1936, at the private meeting of the Military Board of the Supreme Court. The sentence was carried out on the same day. Of course, he was posthumously rehabilitated in view of a complete 'absence of corpus delicti' – let it be a consolation for the defending representatives of 'progressive, leftist intelligentsia' today, because of their much talked-about 'political correctness', dictators, terrorists and hooligans.

Returning to the subject, I can say that, despite a complete absence of any personal connections or acquaintances, I understood from the very beginning, putting it in the words from a popular Russian children's poem, "what is good and what is bad". In short, I decided to specialize in optics because the chair of optics was headed by G.S. Landsberg, one of the people grouped around the remarkable person and physicist, Leonid Mandel'shtam. The associate professor of this chair, Saul Levi, was, luckily, appointed my tutor. I remember him with a warm feeling. I write about him and my relationship with him in the article "Notes of an Amateur Astrophysicist"[2]. There is also some material from this article repeated above and further on; maybe it would be more convenient for readers if everything was repeated here, but

---

[2] This article was published under the number 17 in collection II. Earlier it had been published, though with some small modifications, in the *Annual Review of Astronomy and Astrophysics* **28**, 1 (1990).

I have decided not to do it, for this autobiography is becoming very long as it is. Still, I shall tell how I left optics and experimental physics in general and became a theoretical physicist. When I was waiting, in September 1938, to be conscripted (as a matter of fact, I had already been conscripted and was waiting to be summoned to come with my things to the district military committee), which I again avoided, one can say, as luck would have it (it was the last time when the MGU post-graduates were granted a conscription deferment), I did not want to sit in the room without windows and with the walls painted black, where I was trying to measure the spectrum of radiation of canal rays. So, I made an attempt to explain the effect we were looking for, notably some asymmetry in the radiation of the mentioned canal rays. Notably, I thought that the electromagnetic field of moving charges could cause induced radiation. Such an assumption was erroneous, as the field of a charge is not equivalent to free (light) field. However, I did not understand it all at once and turned with the corresponding question to I. Tamm, who at that time was heading the chair of theoretical physics at the MGU and reading lectures to us. Fortunately, I. Tamm also did not notice at once that my idea was erroneous; he listened to me in a very friendly manner, advised me to look something up, and, in general, supported me. The latter was so important for me, suffering from an inferiority complex and not at all sure of my ability to obtain any theoretical results. I write about all this in more detail in the article "A Scientific Autobiography – an Attempt" published in this book (Article 6). In short, it turned out that I managed, without any complicated mathematics, to sort out some problems of quantum electrodynamics. Specifically, V. Fock, a highly knowledgeable theoretical physicist, and mathematically minded at that, had come to the conclusion when doing quantum calculations that an electron uniformly moving radiates electromagnetic waves. This conclusion surprises, as we in classical theory are used to thinking that a uniformly moving charge (electron) does not radiate. But, the thing was merely to formulate the problem in a different way: in the classical case it is usually a stationary problem, while the quantum calculation had been done with the inclusion of quantum interaction at a certain instant of time $t = 0$. But, the latter is equivalent to the fact that when $t < 0$ there is an electron without the field it entrains, while at $t = 0$ it acquires a velocity $\mathbf{v}$ and starts to interact with the field. Physically, this is equivalent to the assumption that the electron at the moment $t = 0$ is instantly accelerated to a velocity $\mathbf{v}$. It is clear that this process is accompanied by radiation. In general, using the so-called Hamiltonian method, I managed to elucidate some questions and solve a number of electrodynamic problems both in vacuum and in the movement of charges in a medium, for example, the problem of the Vavilov–Cherenkov radiation in the passage of charges through crystals. Here I did not have to use any complicated mathematics. Of course, I found out what had long been known – however close physics and mathematics are, their connection can be very different; in particular, it is possible to make progress in theoretical physics using a very modest mathematical apparatus, for instance, not exceeding the limits of what is taught at the physical departments of universities. In contrast, in some cases theoretical physicists both use and develop most complicated modern mathematics.

Thus, I realized that I could work, achieve results, and think up new possibilities. Such awareness brings a great joy, it is happiness. And I worked a great deal, I wrote a thesis (which is a candidate thesis with us and a Ph.D. thesis according to the Western standards) practically in a year and defended it in 1940. Thus, I finished the post-graduate course at the MGU physical department in two years instead of the regular three; it seems that there it was the first case of such a kind. After that they wanted to leave me, in some capacity, at the MGU Institute of Physics, but the atmosphere there was bad; most fortunately, on 1 September, 1940 I was sent to the FIAN as a person working for the degree of Doctor (such a person has to already have the first scientific degree, Ph.D., and is to prepare, officially in three years, the second thesis for which the D.Sc. is given). I had been doing my postgraduate studies under the guidance of G. Landsberg, as I had been supposed to take up optics. However, he quite nobly, had not hindered me from doing something quite different. Furthermore, I. Tamm, who was considered to be my tutor in my doctor work, also did not hamper my doing what I wanted to do. In general such is the characteristic style of theoretical physicists of the USSR and Russia, at any rate, and undoubtedly at the FIAN.

Thus, I have been doing theoretical physics since 1938, and I have done many dozens of research works on different physical and astrophysical subjects. As for articles, I have already written hundreds of them, because for me writing is fairly easy and writing articles is an integral part of my work. Some colleagues condemned me, and may still be condemning me, for writing such a great number of articles. And indeed, there are people who write articles in order to increase the number of references to their works. But I, believe me, have never been guided by such considerations; besides, I have got very many references as it is. For instance, I have recently seen in a certain reference book that after 1985 there have been 8,962 references to my articles, in spite of the fact that the article which I wrote together with Landau and which was published, in 1950, only in Russian, is almost always mentioned without placing the reference in the literature list; the same source informs that after 1945 there have been 19,519 references to my articles (I began to publish them in 1939). By the by, Landau judged, though not very seriously, the age of physicists by the time when their first publication appeared. For instance, he said that I was 13 years younger than him, because his first article appeared in 1926; and he was born in 1908; therefore, by calendar years I am only 8 years younger. I think that it is not worth attaching too much importance to the number of references; sometimes it is quite deceptive, especially if people refer to some sensational assertion which may prove to be wrong. The number of publications may be slightly more reliable in this respect, but it characterizes the style of working rather than its quality – because it is clear that one publication containing an important result is more substantial

Fig. 5.3. Daughter Irina Vital'evna Dorman with her husband Lev Isaakovich Dorman

than a great number of articles with less important results. Here I will not write about the content of my works, as I can refer readers to the chapter "A Scientific Autobiography: An Attempt" published in this book (Chap. 6). Let others judge the quality of these works, although later I shall make one remark in this respect.

Now I must come back to my biography. In 1937 I married my fellow student, Olga Zamsha. In 1946 we divorced. We have one daughter, Irina Dorman, born in 1939; she graduated from the Physics Department of the Moscow State University, defended her Ph.D. thesis, and was occupied with the history of physics. Her husband, Lev Dorman, is a D.Sc., a specialist in the area of cosmic rays. They have two daughters, Maria Dorman and Viktoria Dorman. The former is not married and lives in Israel. Viktoria has a Ph.D. in physics in Princeton; she is married to M. Petrov, also a physicist, and they have twins, Grigory and Elizaveta, who are now already three years old (they were born in 27 February, 2000). So they are my great-grand-son and great-grand-daughter. They live in Princeton, MT, in the USA.

In 1946 I married for the second time, to Nina Ermakova, who became Nina Ginzburg. Thus, we have been together already for 57 years. Unfortunately, we have no children. As this is not quite an ordinary marriage, at least for the USSR, I have written much about it in a number of articles published in book II. Nina's father, a prominent engineer, was arrested as far back as before the war and died of starvation, in 1942, in a prison of the city of Saratov. It seems that he died in the same cell and almost at the same time as the renowned biologist, N.I. Vavilov. And Nina was arrested in 1944 as a member of a group of young people allegedly going to kill comrade Stalin himself. Of course, it was merely an invention of the KGB, but I am writing about it here, especially as a warning to contemporary 'revolutionaries' and terrorists,

Fig. 5.4. Granddaughter Maria Dorman

so they can foresee their fate in case of their victory. Under a totalitarian regime, the punitive organs deal out justice arbitrarily, and innocent people suffer. Quite often a regular theatre of the absurd adds to that. In Nina's case, she suffered, according to the KGB's scenario, mainly because she lived on Arbat, the street where the great leader sometimes drove by. It was from the window of her flat that he was to be shot at. But the valorous fighters against counter-revolutionaries had not taken the trouble to learn that after the head of the family had been arrested, only one room in their apartment had been left to Nina and her mother, with the windows overlooking the yard. The clearing up of this circumstance, which had happened already after the arrest and ruled out the charge of terrorism, and also some other circumstances, resulted in a sentence surprisingly lenient for that time – 'only' three years in camps. And if her room had overlooked Arbat, we would probably have never met. In 1945, Nina was released by an amnesty, but without the right to live in Moscow and in a number of other cities. Her aunt lived in the city of Nizhni Novgorod (then Gorki), so she went there, in fact, into exile, especially as in Gorki itself she had no right to live as well, and was officially registered in the village of Bor, on the opposite bank of the Volga[3]

In 1945 I was invited by a group of physicists headed by Alexander Andronov, and working at Gorki University, to become a visiting professor at the newly organized radiophysical department of this university. I was 29 and full of energy, but in Moscow I did not have any place to teach (I. Tamm and many others had been actually banished from the MGU by that time). During the war I had worked, among other things, on the propagation of radio waves in the ionosphere. Thus, the invitation to head the chair of propagation and radiation of radio waves was quite natural, and I accepted it. I first came to

---

[3] For foreigners it should be explained that in the USSR a person could live somewhere only with the permission of the militia and had to be registered there.

Fig. 5.5. Granddaughter Viktoria Dorman with her husband Mikhail Petrov and great-grandson twins Elizaveta and Grigory

Gorki at the end of 1945; it was then that I met Nina, and in the summer of 1946 we got married. Of course, I was permanently putting in applications to the KGB (to be more exact, it could be done once a year) with the request to permit my wife to return to Moscow and live where she was born and where her mother lived. However, I was refused all the time, despite the fact that my applications were also signed by the directors of our institute; at first by S.I. Vavilov and then by D.V. Skobel'tsyn. I have written about it in a rather detailed way in "A Story of Two Directors" published in book I. My wife managed to move to Moscow only in 1953, after Stalin's death and the amnesty that followed. It goes without saying that later (in 1956) she was rehabilitated, like all the members of the imaginary counter-revolutionary group – "owing to the absence of corpus delicti". Unfortunately, not all of them lived to see this moment, not to mention the long years of imprisonment for some of those who did.

In 1942 I joined the Communist Party (CPSU). It was just when the Germans reached the Volga, and our view of the future was far from optimistic. Thus, I can hardly be suspected of any careerist considerations, not to mention the fact that I certainly hated the Nazis and all the faults of the communist rule were then receding. At the same time, I cannot help saying, with great regret and bitterness, that for many years I was virtually blind in my estimation of communism and Bolshevism. In general, I believed in "a radiant communist future," not understanding that there we had, in fact, a regime of a Nazi type, headed by a criminal no less mean and bloodthirsty than Hitler. Having written the previous phrase, I remembered the observation of

Fig. 5.6. Nina Ivanovna Ginzburg (Ermakova)

Churchill that Stalin and Hitler differed only in the shape of their mustache. Anyway, I was sharing the lot of millions of people who did not understand the inevitable fate of the totalitarian regime sliding to lawlessness and terror. I have written a lot on this subject (see a number of articles in book II), and it is not possible for me to dwell on it here. I will come back to my own fate.

The years from 1946 to 1953 were far from easy for me. Living mainly in Moscow, I went to Gorky when I could, to read lectures and train a number of postgraduates and workers in physics with an astrophysical and astronomical bias. All that would not have been too bad, but clouds began to gather over my head. State anti-Semitism, and different kinds of persecution of the so-called cosmopolitans, who supposedly worshipped the West, were blossoming in the country. An offensive against modern science was in progress – first of all, in biology, but accusations of idealism and of other sins were raining down on us in physics as well. I proved to be a good target for all sorts of attacks: a member of the party, married to a former prisoner, who had been accused of counter-revolutionary activities and therefore deprived of many rights. And a Jew at that. Finally they started to accuse me of idealism, cosmopolitism and so on. Appearing as a culmination of all that was an article published in a newspaper widely distributed at that time, the *Literaturnaya Gazeta* (Literary Newspaper), on 4 October 1947, that is, on my birthday, under the heading "Against Servility." In the article, inspired by the rather well-known physicist D.D. Ivanenko[4] I was accused of all kinds of sins, together with the biologist A.R. Zhebrak, an opponent of the academician Lysenko. On the same day (what a coincidence!), on the initiative of the same Ivanenko, the High Assessment Commission of the ministry of higher education did not confirm my promotion to the rank of professor, for which I had been nominated by Gorki University. After that, my name began to be mentioned wherever possible as a negative example. Even our institute evidently had to remove me from its academic council to 'strengthen' it. I can only guess what fate awaited me in this situation at that time. I think that it would have cost me dearly, but I was saved by the hydrogen bomb.

The history of creating atomic and hydrogen bombs and my role in it have been extensively described, so I have more reasons to be brief. The Soviet atomic bomb was first exploded on 29th August, 1949, and the hydrogen bomb, on 12th August, 1953. The hydrogen bomb had been arousing interest even before the tests of the atomic bomb, but then it was completely unclear how to make it, nor was it topical, as far as I understand. It was merely something not to be missed in 1948, I. Tamm was connected to this work, though the authorities did not have any special trust in him (once he belonged to the Men'sheviks, who were opponents of the Bolsheviks, and his only brother had been arrested and shot). Only recently have I learned from some jubilee article (the jubilee took place on 12 August, 2003) that at first I had evidently not been admitted to the work in Tamm's group; it only happened some time later. It is a miracle that it happened at all – that someone having a wife in exile was recruited for a top secret job ('top secret, special file'). Of course, I do not know the reasons. I think that the task was not considered very important (see above) and I also had a high 'rating'. However, recruiting my closest friend, and a superb physicist, E.L. Feinberg, turned out to be impossible for Tamm, as Feinberg's wife had once lived in the USA. By the by, I cannot help telling why the services of A. Sakharov were enlisted, though earlier I already wrote about it somewhere, and Sakharov himself mentions it in his "*Memories.*"[5] From 1945 on Sakharov was a postgraduate in the FIAN theoretical department headed by I. Tamm; in 1947 or 1948 he defended his Ph.D. thesis and was going to carry on with peaceful science at the FIAN. He had a small daughter and did not have an apartment of his own, renting a room somewhere, and in general his life was rather hard. So, the director of our institute, S.I. Vavilov, asked Tamm to include Sakharov in his special group in order to get a room for him, which they managed to do (Sakharov got a small room with an area of 14 square meters in a communal apartment). If he had had a room of his own before that, many fates, including both his fate and mine, would have probably turned out quite differently. But

---

[4] As I am mentioning his name, I must say that I am writing about a fact established beyond all doubt (see book II, Article 18, commentary 10 at the end of the article).

[5] A.D. Sakharov. "Vospominaniya" (Memories). Moscow: "Prava Cheloveka" (1996). There is also an English edition.

it happened as it happened. From the middle of 1948 we started working, at first reading the materials which we had and which, as far as I remember, did not raise any hopes that this problem would be solved. Then Sakharov put forward "the first idea" and I "the second idea", which together made it possible to build the first Soviet hydrogen bomb. It is ridiculous, but up until the death of Sakharov in 1989, that is, 40 years later, all this material had been classified. That is why Sakharov, in his memories, writes about the 1st and the 2nd ideas without revealing their content. Now it has long been known that the 1st idea is "the layer structure" (I will not explain it here), and my 2nd idea is using as fuel ^{6}Li (or, if you wish, ^{6}LiD) for obtaining tritium ^{3}H $\equiv$ t as a result of the reaction ^{6}Li $+$ n $\rightarrow$ t $+ ^4$He $+ 4.6$ MeV. These two 'ideas' were recognized as opening the possibility of creating the hydrogen bomb. With the aim of carrying out this task, Tamm and Sakharov were sent, in 1950, to 'the site' (Arzamas-16), and I was not admitted there, obviously, for the aforementioned reasons (it concerned a higher level of secrecy), and I remained in Moscow as the head of a small 'group of support', but with a sentinel sitting at our door. It goes without saying that I regard this turn of the fate as the greatest luck. I was sure of my safety, so I could visit my wife in Gorki, and could do science. It was not that I was neglecting my work, but our task was mainly performing different calculations, which is not my strong suit. Taking an interest in other people's secrets was not customary with us in general, and I was especially indifferent to them. Thus, I remember my surprise, probably, already in the 1970s or the 1980s, when by chance I asked Sakharov how 'the layer structure' had 'worked' or developed. He said that it had 'not worked', which I did not understand. Now it is known that the first two Soviet hydrogen bombs were made as 'layer structures'. But such a construction proved to be able to increase the power of the bombs, I do not remember exactly, whether 20 or some other, comparatively small number of times more than that of the bomb dropped on Hiroshima. Somebody thought for some reason that it was not enough, and the 3rd idea (by Sakharov's terminology) consisting of radiative compression was used. While 'my' 2nd idea is probably still working. Or rather, not working, to our great luck, but merely figuring.

Trying to be worth my 'classified' salt, as it has already been said, I was glad when Tamm or Sakharov (I do not remember exactly which one), having come to Moscow in 1950, told me of the problem of controlled thermonuclear synthesis and the "Tokamak" system offered by them. I took up this problem, managed to write several reports and, by the by, in 1962, when all that was declassified, even published some material from these reports. But I did it as a sort of some rather silly 'revenge' for having been dismissed from this work in 1951, which must have been considered too secret for me (for more details see book II, Article 18).

Then an especially terrible time came. Stalin went totally insane, repressions were going on, culminating in 'the case of doctors' and a bestial related anti-Semitism; it seems that the corresponding documents have not been found yet. They may not have existed, as it was clear even to bandits that it was better not to leave a trace. According to the rumors, 'the doctor killers' were either to be hanged in the Red Square or exterminated in some other way, and all Jews were to be exiled into some camps that were already built. Some 'necessary' people, possibly myself among them, would have been left in 'sharashkas,' later described by Solzhenitsyn and others (a 'sharashka' is, in fact, a prison were scientific and technical research was conducted). It was tremendous luck that the Great Leader did not have enough time to carry out what he had planned to do and died, or was killed, on 5 March, 1953. In the former USSR many people (at any rate, my wife and I) have up till now been celebrating this day as a great festival.

Everything in the country after Stalin started to change very quickly; suffices to mention the rehabilitation of 'the doctor killers' and the shooting of Beriya, who was the head of the Soviet 'atomic project' (by the by, he was a good organizer and probably not more of a bandit than all the rest). At the USSR Academy of Sciences, contrary to its charter, there had been no election since 1946, evidently because it had not been allowed by Stalin. In 1953 such an election took place, and I was elected a corresponding member of the Academy. As they say, 'from rags to riches' – although, in my case, this saying looks somewhat exaggerated. But still, both in the USSR, and now, in Russia, being a member of the Russian Academy of Sciences means a rather privileged position. In addition, I got some rather high governmental awards[6] and, in general, turned into a VIP, although second-rate. No less important was that my wife managed to return to Moscow, after 8 years of half-exile, not to mention approximately a year and a half of prison and camp.

Since that time life has been going on more or less normally, but still it makes you think of a popular joke: "Question: What is permanent under the Soviet power? Answer: temporary difficulties." The Sakharov story became a temporary difficulty for me. In 1969, dismissed from classified work, he returned to our theoretical department of the FIAN. As for me, in 1971, after Tamm's death, I had to become head of this department. 'Had to become' indeed, because I did not want it and in general I do not at all like work like that. However, the workers of the department had asked me, and it was really

---

[6] I must say I do not feel like enumerating all these awards, including prizes and being elected members of some academies. Enumerating all of them would be too long, and mentioning only the main ones would mean a lack of respect for others. The corresponding information. although incomplete, can be found in books I and II. Let me note that I have never strived for awards, and never aspired for them, though I was glad to get them. Besides, the Nobel Prize has exceeded everything. I know that they had started to nominate me for this prize as far back as about thirty years ago. However, I had long ago come to the conclusion that it had been decided not to give it to me. By the by, I was able to understand it – both because of the certain circumstances of my life and because I knew that the work of the Nobel Committee is extremely difficult. That is why the news of the prize having been awarded to me, in 2003, was quite unexpected.

necessary, because due to numerous 'temporary difficulties' it was important that the department should be headed by a person of some rank. And in our department only Sakharov and I were academicians (I had been elected a full member of the Academy in 1966). However, Sakharov had already become a dissident, had taken up politics, and was totally unsuitable for the post. So, I had to take over. And it was the same Sakharov who first of all became a problem, one can say, especially after his being exiled to Gorki in 1980. I have written about it in the article "The Sakharov Phenomenon", which can be found both in book I and in the book of memories about Sakharov. I will not repeat myself; I will only say that I do not see anything for which I could reproach myself in this matter.

However, I must note that 'the case' of Sakharov aggravated my already not temporary but permanent problem, which I faced when under the Soviet regime; I either would not be allowed to go abroad at all or would be allowed with great difficulty. The explanations always referred to the reasons of our famous 'secrecy', but it was an obvious pretext. For instance, I. Tamm, who really did know some secrets (while I actually knew none), went abroad many times after 1953. Of course, I was very glad for him. But I was refused permission to go, because of the 'sins' of my wife and my own sins. My greatest 'achievement' in this area was made in 1959. That year, a big international physical conference took place in the USSR, in Kiev; it was called the Rochester conference. Some people, myself included, were not allowed to go to this conference (!) under the pretext of secrecy. Landau was also not allowed at first, but he announced that he would go all the same and would make a scandal. This was well done; they retreated and he went there. I cannot make scandals and, even if I could, I would perhaps not have obtained anything by it. So I did not go to this conference, did not see many good physicists, and did not learn many interesting things. I was very bitter about it at that time, and I am bitter now when I remember this insult. And the greatest pity is that I could not meet Einstein and speak with him – for he died in 1955 when I was already 39, and understood something in physics and astronomy already and could well have had a professional talk with him if I had lived in a free world. But I was not fated to do this. I only remember asking L. Infeld, who came to the USSR, to give my regards to Einstein. I do not know if Infeld had an opportunity to do it.

Going on with the subject, I shall describe two later experiences characterizing the conditions in which we lived before the collapse of the USSR. In 1984 the Danish Academy of Sciences, of which I have been a foreign member since 1977, invited me to come to Copenhagen for a week; I do not remember on what occasion. So I started the painful procedure of 'registration'. For going abroad one had to fill out many papers and pass some commissions. Then all the documents went to the 'instance', and sometimes you were not informed until the very last moment whether or not your trip had been permitted. This time I was told shortly before the departure that I could go, but without my wife (!). I refused to go. Such a natural reaction was a rare thing at that time, so I even had a telephone call from the president of the USSR Academy of Sciences, who expressed his disapproval by saying that he, for example, did travel without his wife – so how could I be so shrewish instead of being grateful for the great honor done to me and the trust given (to be honest, all the last words were not said but, in my opinion, this call could have been interpreted only in such a way). A year later, in 1985, again in Copenhagen, Niels Bohr's birth centenary was to be celebrated. And again I was invited, this time, if my memory does not fail me, being the only person from the USSR who was to make a report at a plenary meeting of the corresponding conference. Again, I started to go through the 'registration' with my wife. And again at the last moment I was told that I was allowed to go, but without my wife. Probably, someone was afraid that if we went together, we might not return (by the by, we had never thought of anything of the kind). However, this time I went, though I was indignant. The thing was that I had already prepared my report, which was quite a lot of work, taking into account that I had to prepare it both in Russian and in English and then to pass it through censorship[7] And, what was even more important, I certainly had a great respect for Bohr (I had heard his reports and was acquainted with him, because he had been to the USSR) and I wanted to take part in the jubilee conference. In my report at the conference, and especially in private talks, I somehow expressed my indignation at our lack of freedom. I do not remember the details, but I do remember the bitter aftertaste left by the indifference of Western colleagues. However, I was probably wrong; in the West they knew about the Soviet arbitrariness and were used to it, they protested and defended some people. And in general, why should they have taken my case to heart? Besides, I spoke in a reserved manner, without any attempts to arrange a rally, so it was easier to turn a deaf ear.

By a chance coincidence, this year, 1985, was a turning point, as M.S. Gorbachev came to power, and soon the 'perestroika' began. I took part in it – although not very actively, but I did. I was a deputy of the USSR Supreme Soviet (from 1989 to 1991, the year when it was dissolved), elected from the Academy of Sciences. I resigned from the party in 1991, and since then I have never again been a member of any party and am not going to be one. Of course, I have always been a partisan of the democratic forces, but I have not always identified myself with the concrete actions of some of their representatives, Sakharov in particular. However, I do not remember anything important and interesting from my 'political' activities, which would be worth describing here.

In 1988 I finally managed to free myself from heading the I.E. Tamm Department of Theoretical Physics, as it began to be called. The thing is, that according to the rule which had just been introduced, and which had long been necessary, people over 70 were not allowed to hold some positions. Having remained at FIAN in the position of advisor to the Academy of Sciences, I have

---

[7] Fortunately, now things are quite different!

a small group of subordinates. Besides, I have been heading, since 1968, the chair of problems of physics and astrophysics, created that year at the Moscow Physical and Technical Institute (rather well-known as Phystech), but now I do not read lectures and remain in this position without pay (according to my own wish), at the request of the staff in order to be able to help the chair in solving some problems.

Since 1998, I have been chief editor of the *Usp. Fiz. Nauk* (Physics–Uspekhi) journal, and I actually carry out my responsibility to the journal as far as I can. From the middle of the 1950s I was in charge of a physical seminar at the FIAN, on Wednesdays, and this seminar was rather popular. Running for two hours, it was attended by many people from Moscow and from other places as well. At the time of Khrushchev we had a popular saying: one should know the measure, be it maize or seeing Nehru with pleasure. The saying came into being in connection with Khrushchev's keenness on cultivating maize even in the areas totally unsuitable for this purpose, as well as with his love for Javaharlal Nehru (which the latter may have deserved). I am writing about that by association. In 2001 I started to feel worse and decided that I "should know the measure." Therefore, on 21st November 2001, the 1700th seminar[8] was arranged in the form of an amusing party, like some other 'jubilee' seminars (for instance, the 1500th and 1600th), and there, quite unexpectedly for everybody, I announced that the seminar was closing down. I did it after telling a story about an actress who had been 'playing' until she could no longer walk. I still could walk, but the seminar had to be closed some day and I wanted to do it in time.

In post-Soviet Russia, and even a bit earlier (after 1986 or 1987) going abroad was no longer a problem, or at least no longer a problem of the Soviet type, that is, with a lot of obstacles. I took advantage of the corresponding opportunities, but due to 'the law of conservation' going abroad has become difficult for other reasons – because of age. In 2001, I was unexpectedly awarded a Humboldt fellowship which meant going to Germany for half a year. I was already going to accept this invitation, but at the last moment refused it because of the state of my health. However, I used the visa already received for going to Spain to the 10th International Conference on Ferroelectricity, the first conference of such a kind which I managed to attend (for more information, see book II, Article 5). I thought that I would not go abroad any more, but here I am, going to Stockholm to the Nobel award ceremony.

A great deal has already been written by myself already, but nevertheless, I do not think that some more questions could be left without answers. Indeed, if we want our notion of a person to be full enough, we should learn many things about him. Specifically, we would like to know, firstly, what his outlook is, including his attitude towards religion. Secondly, we would be in-

---

[8] For more details, see the book *The Seminar: Papers and Reports.* (Compilers: B.M. Bolotovskii, Y.M. Bruk) (Moscow: Izd. Fiziko-Matematicheskoi Literatury, 2006).

terested in his political views. What follows, thirdly, would be the description of his professional activities. And finally, fourthly, importantly, what are his personal qualities and tastes. All that should be described. But of course, it is easier wished than done. Besides, I think that there are many personal things about which a man sometimes cannot and should not write. Indeed, in newspapers and magazines we can often see the answers to such questions asked by correspondents: what qualities in people do you value above all else, what are your ideals, do you want to earn more money and for what purposes, and so forth. The answers, as far as I could observe, are rather monotonous: everybody values faithfulness, nobleness and decency; everybody has radiant ideals of fairness, and money is needed mainly for good deeds and not for buying mansions. It is quite understandable, as everybody knows, what qualities and likings are considered positive and what tastes one should and can be ashamed of. So I will not go into details of my personal matters. I will only say a few words on the theme related to them. In my life, like in the lives of many other people, friends have played an important role. Now some of them are already no more, to my greatest regret. Two closest friends of mine are now having hardships connected with old age. I would like at least to mention all my friends and to express my warm feelings for them. However, after some attempts, I have seen that I am not able to do it in a satisfactory way. The same concerns my wife and relatives. Probably only a few people will be able to write about their feelings quite sincerely and without hitting any wrong notes.

One more category of person to be described in a complete enough autobiography are teachers and pupils. Treating the matter seriously, it is not at all simple to decide whom you should regard as your teachers and pupils. Evidently each of us have learned a lot from many people, in person or externally. Where I am concerned, I want to solve this problem as I.E. Tamm and L.D. Landau solved it. The latter said, quite definitely and repeatedly, that it was only Niels Bohr whom he considered his teacher. As for I.E. Tamm, as far as I know, he regarded only L.I. Mandel'shtam as his teacher. For my part, acting in the same way, I believe that my teachers are I.E. Tamm and L.D. Landau. By the by, with the latter I did not have any official links – I did not take an exam in his 'theoretical minimum', he was not my tutor or superior, nor did his name appear as that. About Tamm and Landau I have written in a rather detailed way (see books I and II). As for the pupils, many people (especially mathematicians) call their pupils those whom they either supervised, for instance, in their postgraduate studies, or for whom they were considered tutors helping to prepare a thesis. I cannot be satisfied with such an approach. I want to call my pupils, although maybe not always, those who consider me their teacher. Naturally, with such an approach I cannot present a list of my pupils. I will only say that I have never forgotten the role played through the support and benevolent attitude of I.E. Tamm. I tried to follow his example. Let other people judge to what extent I have succeeded in it. I will confine myself to the remark that, with only very few exceptions, I have good feelings about those with whom I either happened to work or whose tutor I used to be. Thanking them for cooperation, I hope that they returned and return my feelings.

Where my professional activities are concerned, they are described in the Nobel Lecture, in this autobiography and, the main thing, in a number of articles in books I and II; I would especially like to note the article "A Scientific Autobiography: An Attempt" in book I (Chap. 6 in this book). It is interesting to consider the question of the estimation of people's scientific level and activities. It is a big theme and here I only want to take an opportunity to tell about Landau opinion in this respect. Landau, in general, especially in his youth, liked to classify everything. Later, when I got to know him, he himself looked at these classifications, for instance, of women, with irony. I shall only dwell on his classification of theoretical physicists who worked in the 20th century. It was done according to a logarithmic scale, with the base 10, so a 2nd class physicist was 10 times inferior to a 1st class physicist. It was a question of accomplishments, and not of the level of knowledge, of the pedagogical ability or the oratorical talent. Class 0.5 was given only to Einstein. Put into class 1 were Bohr, Dirac, Heisenberg, Schrödinger, de Broglie, Feynman. I have undoubtedly forgotten several names, but, of course, I will not supplement this list in the way I see it. I remember Pauli being put into class 1.5. Landau himself was placed in his own classification, as far as I remember, at first into class 2.5, and then to class 2. Once E. Lifshits told me that Landau had upgraded himself to class 1.5. Why does this classification seem interesting to me? Firstly, it refutes the legends about Landau's immodesty and conceit. Secondly, it was important to him that he put an emphasis on a record of accomplishments and achievements. For instance, I remember arguing with Landau about de Broglie, who did not seem to me a particularly powerful figure. But Landau stood firm, and I think he was right. The wave properties of the electron and other particles was a guess of genius, although this idea occurred to de Broglie under the influence of Einstein's notion of photons and the equations $\mathbf{p} = \hbar\mathbf{k}$ and $k = \omega/c$. By the by, I am surprised that Landau put Feynman higher than himself and, in general, put him into the 1st class. There is no doubt that Feynman was a brilliant physicist and lecturer but it seems to me that his accomplishments cannot be compared with those of other 'first-class' physicists. Probably, Landau especially valued the diagrammatic technique, thinking that he himself would not have been able to hit upon it. What looks interesting and important, from my point of view, is an enormous gap which may exist and often exists between accomplishments and knowledge, the command of the apparatus and so on. I think, though I cannot assert it, of course, that 'by knowledge and command of the apparatus' Landau stood even higher than Einstein, but 'by accomplishments' he sober-mindedly estimated himself as much inferior to Einstein. I am certainly not going to give my estimation of myself here, but it is undoubtedly much lower 'by command of the apparatus' than 'by accomplishments'.

It seems to me that working in science, in physics – at any rate, in a position of at least some authority – is absolutely impossible without having a stand on the important issues of our existence and thinking about philosophical questions. Unfortunately, I have not had the chance to really acquaint myself with the philosophy and methodology of science. To a certain extent it can be explained by the fact that I have always worked at concrete tasks and not at, for instance, problems of the interpretation of quantum mechanics and so on. That is why the atmosphere of dogmatism and censorship in philosophical issues, which prevailed in our country, has not had an especially pernicious effect on me, as well as on the majority of my colleagues in the USSR. I do not even want to name illiterate rogues who dictated to us, under the guise of dialectical materialism, how the laws of physics or, for instance, of genetics should be understood. However, this does not in the least prove that dialectical materialism in itself is unacceptable (among its advocates there were quite decent and reasonable people, as for instance, the already mentioned B. Gessen). I have always understood the core of this philosophical trend as, firstly, materialism; that is, the conviction that nature and matter exist beyond our consciousness and independently of it. Secondly, studying nature, which is what natural sciences do, requires a flexible approach based on facts and not on dogmas. It is such an approach which is dialectical. I do not mean anything else, although philosophers may consider my words naive, insufficient, or maybe even worse[9] A really important point is understanding that materialism is 'an intuitive judgment' (using the term especially widely used by E.L. Feinberg) which can be neither proved nor disproved. Atheism and the belief in the existence of God are also intuitive judgments. I am an atheist, that is, I think nothing exists except and beyond nature. Within the limits of my, undoubtedly insufficient, knowledge of the history of philosophy, I do not see, in fact, any difference between atheism and the pantheism of Spinoza. That is why I think that Einstein was also an atheist, because in 1929, when asked what he believed in, he answered: "I believe in Spinoza's God, who shows himself in the harmony of all that exists, but not in a God who takes care of the fate and actions of people." Einstein, however, used the term 'cosmic religion' and reckoning him among atheists may be not quite right. At any rate, I, like many others, do not have any 'cosmic feeling', and I do not see any place for God, that is, for something which is beyond nature and has created this nature (such is the opinion of deists). But, evidently it is impossible to prove that God does not exist. This itself leads to the conclusion that the principle of the freedom of conscience is fair, that is, people should have the unimpeded right to believe in God and, if they so desire, to practice

---

[9] As it was explained to me, a more exact interpretation of dialectical materialism is seeing it as a combination of materialism with the recognition of the laws of dialectics formulated by Hegel, whereas I regard these laws as scholastic to a considerable degree. Perhaps the most correct name for the views that I hold would be scientific materialism.

some religion (of course, I am not speaking about wildly fanatical sects and the beliefs justifying banditry and terrorism). The Bolshevik communists were not merely atheists but, according to Lenin's terminology, militant atheists. This term, being not quite clear for the masses, was changed to the name 'the militant godless'. These pursued believers, especially priests, destroyed temples (churches, mosques, synagogues) turning some of them into warehouses, stables and so on. Identifying atheists with 'the militant godless', which is done in Russia either by unscrupulous people or simply by illiterate demagogues, is absolutely groundless. It is the same as identifying a respectable Catholic with a partisan of the Inquisition or considering all members of the Orthodox Church to be partisans of the brutal persecution of Old Believers or other 'heretics'. Unfortunately, in the post-Soviet time in Russia, a clerical offensive has been going on, while the voice of atheists is completely stifled. That is why since 1998 I have been defending atheism in the press, and after being awarded the Nobel Prize I managed to talk about that on television as well. Here it is not a question of combatting religion, as this would be at variance with the principle of the freedom of conscience, and, besides, religion sometimes does good, calling for good and for worthy behaviour. It is merely a question of atheistic enlightenment, for instance, of scientific elucidation of a complete falsehood of creationism. In general, an abstract belief in the existence of God (concretely, deism) should not be confused with theism (Christianity, Islam, Judaism) involving a belief in miracles and the sanctity of the Bible, the Koran and so on. By the by, no cultured person will deny the great artistic and historical value of the Bible. But a belief in Biblical miracles is another thing altogether, being incompatible with a scientific outlook, because a miracle, by definition, is something at variance with scientific data and the results of scientific research (for some more details see a number of articles in book II).

It remains for me to write about politics, or rather about my opinion on certain questions in this respect. I have always remembered the line from a song by the popular poet and singer Alexander Galich (by the by, his real surname was Ginzburg but we are not relatives): 'Be only afraid of the one who will say: I know what should be done.' Lenin, Hitler and Stalin knew 'what should be done' and this knowledge cost the lives of millions of people. Even if I knew 'what should be done', I would not be able to influence the course of history. But the main thing is that I do not know this. I have some opinions, but there is a world of difference between being sure that you are right and acting correspondingly, in spite of everything – and having an opinion and understanding that it may prove to be wrong. So, my opinion is well-known enough: I am convinced that only a democratic form of rule is acceptable. Churchill was certainly right when he said, with the clarity of thought inherent in him, that (unfortunately, I do not remember his exact words) that democracy is a very bad form of rule but we do not know any better. Indeed, a hereditary monarchy is presently something which remains merely by tradition. Even so, the monarchic regimes remaining in some countries of Europe do not hamper the democratic order, as far as I know. Where the totalitarian regime is concerned, Lenin, Stalin, Hitler, Mao Tse-tung, and many leaders of a lower rank proved convincingly enough that this form of rule is inhuman and inevitably leads to tragic circumstances. Millions of people, and I am not an exception, believed, putting it in a more or less modern language, in the possibility of 'a communism with a human face.' To tell the truth, I still do not understand why it is impossible. But there is 'a criterion of practice', in exact sciences which is absolutely unshakeable and enables us to have the indisputable knowledge, for instance, that astrological forecasts are false and the law of conservation of energy is correct (though for the latter this is with a reservation about the whole known range of phenomena). In social life there is certainly no similar indisputability, but the whole experience of mankind shows that totalitarianism and dictatorship inevitably slide into the rails of arbitrary rule, crimes and atrocities. However, for democracy a high price is to be paid too, and arguments emerge about this price – how far can we go, and where is the border? The whole history of the rule of the Bolsheviks and the Nazis is full of examples of how democrats, progressive intellectuals, liberals, pacifists and such often did not understand elementary things, and step by step they in actual fact encouraged bandits and consolidated their power. Much has been written on this theme, and I could also write a whole book. I will confine myself to just one example of the 'deep understanding' of Stalinism, which I have recently come across. Herbert Wells wrote, characterizing Stalin, that he had never met a person more sincere, decent and honest; that in Stalin there was nothing dark and sinister, and it was these qualities that should account for his tremendous power in Russia(!). According to Wells, Stalin was a Georgian, totally devoid of cunning and insidiousness, and his sincere orthodoxy guaranteed the safety of his comrades-in-arms. But then, it was written in the 1930s, already after the horrors of Stalin's collectivization, known all over the world. Several years later Stalin's comrades-in-arms, who Wells had probably met, were also shot. In all, the 'Georgian devoid of cunning and insidiousness' signed by his own hand death sentences (shooting lists) for 44,000 people[10]!

There are really lots of examples of 'understanding' reality in the way demonstrated above. However, I have an impression that the 'leftist progressive' public 'has forgotten nothing and has learned nothing.' Under the slogans of 'political correctness,' following democratic principles and so on, they follow the same road of protection and justification of hooliganism, banditry, and terror. Even the tragic events of 11th September, 2001 were far from having a sobering effect on each and everyone. A spectacular example is the attitude to the events in Iraq. The Iraqi regime of Saddam Hussein used chemical weapons

---

[10] See A. Yakovlev. "Sumerki" (Twilight). Moscow: "Materik", 2003. The German translation has already been published, the English one is being prepared. The citation from H. Wells is made by re-translating the Russian text in the Russian edition of book II in comment 12* to Article 25.

in the war with Iran, it occupied and plundered Kuwait, and it committed innumerable atrocities inside the country. But they say that it has not been proved whether Saddam did or did not manage to build up a stock of chemical, bacteriological, and maybe also nuclear weapons. It has been proved, however, that he aspired to that. All the vileness of this regime and its crimes have also been proved. However, the actions of the USA and Britain taken in order to overthrow this regime is condemned by many people. In their opinion, they should have waited until Saddam became stronger and was able to drop something on their heads. I understand that the Anglo-American action was clumsy, and this issue is complicated. However, I do not understand why the elimination of the regime of Talibs in Afghanistan had been considered more legitimate from a formal point of view. It is just that 11 September was closer, and bin Laden and Afghanistan farther. I have already said that I do not know 'what should be done.' But I know, as it seems to me, that if civilized countries follow the principles of 'political correctness' and observing laws literally, they will again incur innumerable calamities, like those from Hitler and Stalin. The wild fanatics killing absolutely innocent people, women, and children, cannot be stopped and rendered harmless without soiling one's hands.

As a Jew I cannot go on without dwelling on the 'Jewish question' here, although it is still not entirely clear to me why this quite small, suffering group of people turned out to be in the focus of world politics. For an understanding of what follows I must make several observations. The Jews have remained as a nation, and they have not assimilated, as it is customary to think, due to their attachment to Judaism and because they consider Israel a chosen nation. I am an atheist and internationalist, that is, I do not regard any people as 'chosen'; in particular, I do not think that Jews are better than Arabs. I do not know either Jewish language (Hebrew or Yiddish); by the by, I wish I knew them, but I do not have the ability for languages and my native tongue is Russian. One would think that I must have assimilated. But it is absolutely wrong; I have never been able to even think of giving up my native people. What are the reasons? I do not know and understand them quite well myself. Of course, family roots are essential; there were some Jewish traditions in my family. No less essential is anti-Semitism. Although I have not suffered from it directly, I also used to be called 'a Yid', not to mention the time when after the war anti-Semitism flourished in the USSR. Anyway, because of all these reasons and maybe some other reasons which I do not understand (it may merely be a question of genes, and their role is not quite clear yet), I am a bearer, so to say, of the Jewish national feeling. This is by no means nationalism, for I see nationalism as the opinion of the superiority of 'one's own' nation or, at least, striving to justify and defend 'one's own' people. I categorically deny having such feelings. On the contrary; with me the most spectacular display of the Jewish national feeling are shame and indignation when I face a Jew who is a scoundrel and on the whole a bad person. At the same time I am glad if a worthy person turns out to be a Jew. For instance, I am happy that Einstein was a Jew, as well as many other outstanding people. In these feelings I see nothing to be ashamed of. It is shameful to promote somebody who is of 'one's own' at someone else's expense, and it is shameful to pardon a scoundrel who is one of 'one's own'. Taking pride in 'one's own' good person is not shameful to me; however, I cannot clearly explain and understand these feelings of mine. But this is another question. By the by, to avoid a suspicion of being insincere, let me say that my first and my second wives are Russian. My daughter's husband is a Jew. My granddaughter's husband is Russian. I do not see any problem here; life is like that.

I have burst out with the above partly in order to explain why I was interested in Israel and its fate. I am very glad that there is such a state where at last the Jews are not a minority, often persecuted and humiliated. At the same time, there are many things there I do not like. For example, and this is not in order of importance, I cannot help noting clericalism. In the history of the Jews, synagogues played not only the roles of prayer houses but also the roles of the centers of a communities. So, some state support of religion in Israel can be understood. However, it is important to know the measure. I cannot see any reasons justifying the absence (at least, in a number of places) of public transport on Saturdays and a number of other, sometimes more important, restrictions or consequences of a religious character. In such a way, atheists are discriminated against. Still more important are a complete absence of unity in the country and the presence of a great number of cases of abusing the advantages of democracy, which I have mentioned above. That is why, I think, the very existence of the state of Israel is under threat, while the elimination of this state would be a new catastrophe of the Holocaust sort. I am writing about it here, although I am aware that it might be irrelevant, as I am indignant at the support of Arafat and his gangs by anti-Semitic, and at the same-time progressive, 'ultra-left' forces of the West. And it is not because of my supposedly anti-Arab stand. Stalin and Hitler would have solved the Palestinian problem in 48 hours. They would have either eliminated Israel or, which is much less likely, the Palestinian autonomy, deporting the disagreeable population to some distant place. Of course, in the civilized world such a decision is inadmissible. I am of the opinion that it is necessary and possible to have two completely isolated states. All those who either know the lessons of history (for instance, the history of the 'friendship' of Catholics and Protestants in Northern Ireland) or who have seen on television dancing crowds of Palestinians of all ages while the television was showing the terrorist acts of 11th September, which took away the lives of thousands of innocent people, cannot have any illusions about the friendship and love between Palestinians and Israelis. To be more exact, only absolutely brainless bankrupts who had once obtained the celebrated agreements in Oslo can assume (or pretend to assume) that this might be possible in the foreseeable future. That is why I am convinced that a friendly coexistence of Israel and the state of Palestine is nowadays impossible. The only way out, in my view, is a complete isolation of these states from each other. It cannot be done with the majority of Israeli 'settlements' remaining in the Palestinian territory. They should be removed, and it is not a question of law. One Israeli man told me that the land for these 'settlements' had been bought and not annexed. So much the better, but it merely means that they can try to 'exchange' these settlements for the land of the Arabs living in Israel. These Israeli Arabs are the only true obstacle to a complete division of the two states.[11] However, this obstacle is not decisive either. Of course, an impenetrable 'wall' dividing the two states is also necessary. Using Palestinian workers in the territory of Israel is absolutely inadmissible. This is motivated, as I heard, by humanist considerations, allegedly by caring about poor Arabs, while in actual fact, I think, many Israelis do not want to do manual or menial dirty work, and their own workers are, in spite of unemployment, either more expensive or unavailable. But the Palestinians working in Israel can't help hating their rich masters, which is a source of additional antagonism. The history of the importation of black slaves to America could be a good lesson. And taking care of Palestinians should be carried out by their brothers in rich Arab countries.

As for the question of the Golan Heights, it seems to me completely made up. I have been there myself and seen the ruins of a big ancient synagogue. So why is it Syrian land 'from time immemorial'? Syria attacked Israel, suffered a crushing defeat, and lost the Golan Heights. Now there is no Syrian population in this territory, and its loss is Syria's pay for aggression. Why do those who are indignant at this situation not demand that Konigsberg and its environs, renamed Kaliningrad and the Kaliningrad region, should be returned to Germany? Such is a result of Germany's attack on the USSR, and nobody, including Germany, is going to revise it. By the way, I am indignant that a part of Eastern Prussia and Konigsberg were named after Kalinin, a nonentity, who licked Stalin's boots, while Stalin sent his completely innocent wife to a concentration camp.

Being a realist, I am sure that such a 'Ginzburg plan' will not be realized and will only arouse the spite and mockery of pseudo-democrats and 'peacemakers'. That is their business, while I want only to express my opinion, taking advantage of freedom of speech.

In conclusion, one observation of general character. A tremendous progress of science has led to its deep internationalization. There is no such thing as American, Russian, Jewish or whatever else national physics. There is only one physics in the world, and when we are speaking, for instance, about British or Russian physics, we only mean the organization or the state of physics in Britain or in Russia. The so-called Aryan science of the Nazis and the Marxist and Leninist science of the communists have long been forgotten. Whereas, in the field of the social sciences and the sciences related to them, such as

---

[11] There also is a certain problem because Jerusalem is a very important center from a religious point of view, not only for the Jews but also for the Christians and the Moslems. However, I do not see why a free religious life cannot be provided under a correspondent agreement (and, perhaps, also under international control) and in the conditions when Jerusalem is the capital of only Israel.

sociology, psychology, economics and so on, true depth and internationalization are still far from being achieved. However, I hope, there also the time of great success is near; such is one of the factors which enable us, as it seems to me, to look into the future with hope. Another factor is the slowdown of the growth of the population of the Earth. Also positive, though this assertion is disputable, is the increase of the average life expectancy and thus the increase of the average age of the population. I remember the saying: if a man is not a communist at the age of 20, it means that he has no heart, but if he is a communist at 50, it means that he has no head. The bitter statement 'what history teaches is merely that it does not teach anything' is not without reason. But, still, to completely agree with this statement would mean to totally lose faith in mankind.

What has been said explains why I am still inclined to believe in the radiant future of mankind. Today, on the road to it, there are many obstacles; the Islamic (terrorist) threat, poverty and the lack of education of the great masses of population, AIDS, and other diseases. But, let us remember the situation, for example, in 1943, sixty years ago. Europe was under Hitler's heel, and the USSR, though heroically resisting, was living under the Stalinist yoke. America was not so strong, and a world war was raging. Was it easier and better than now? The forces of democracy have coped, and saved civilized society, and nowadays both Nazism and communism have almost sunk into oblivion. That is why we can hope for the ultimate triumph of the democratic system and secular humanism[12] all over the world. The necessary conditions for both are the presence of historical memory and the development of science.

22nd November, 2003

Notes on the English Translation

This autobiography was written in November, 2003, when I had not yet left for Stockholm for the Nobel Prize ceremony.

This ceremony, and all the other corresponding activities in my case, probably went in a standard way, and I do not see any reason to describe them. Perhaps I will only mention that, when at one of the banquets I had to respond to congratulations, I advanced a theorem: "Every physicist will get the Nobel Prize if he lives long enough."

It's a shame that many people worthy of the prize just did not live long enough to win it. We should bear it in mind, which in itself is a good enough reason not to overestimate the place and importance of the Nobel Prize. As is known, the information on who nominated whom for the prize is made public only 50 years after the prise is awarded. Thus, in my case, the information will not be available until 2054. I heard, however, that I was first nominated

---

[12] For more on secular humanism, see Article 35 in II (see the footnote on p. 135).

as far back as the 1960s, and repeatedly nominated since then. In almost all such cases known to me, the prize was to be given for the GLAG works, whose authors were Ginzburg, Abrikosov and Gorkov (the letter L in GLAG corresponds to Landau, who passed away in 1968 and was awarded the prize alone in 1962, the year when he was involved in a car crash).

Since I had not been given the prize for many years, I thought that I would never get it. I did not in the least take offence, for I did not have megalomania and did not think that I was bound to be given the prize. I thought that I could only be given it without detriment to its dignity and average level. And, by the way, in my opinion, that is the case of the majority of the prize winners. If there are good reasons for 'could be given' to, say, 15 or 20 people out of the 300 nominated, and out of them not more than three should be chosen, it is clear how difficult is the task of the Nobel Prize Committee. So, having assumed that I would, most probably, never get the prize, I decided, when I reached the age of 80, to sum up the results of my work in the field of superconductivity and superfluidity. That makes the second article in the present book. The existence of this article brought about the not quite standard form of my Nobel lecture (see the first chapter in this book).

When I came back from Stockholm (there, and on the way home, in Helsinki, I delivered several more lectures), I found myself facing a barrage of mass media, and television, and newspapers in particular. The favorite subject of the questions was: why do they 'neglect' the USSR and Russia – rarely giving their scientists the Nobel Prizes, especially in comparison with the USA? I do not presume to judge about all prizes, but in the case of the prizes for physics I have quite a definite opinion, and I think it would be worthwhile to give it here. Namely, nobody neglects us, except perhaps ourselves. There is a lot of corresponding data in the book by A.M. Blokh[13]. Before the October upheaval, that is, in Tsarist Russia, the only nominee for the prize for physics was P.N. Lebedev, who was the first to measure the pressure of light. He was nominated for the Nobel Prize for Physics twice (in 1905 and 1912), and there is almost no doubt that eventually he would have gotten it. However, Lebedev passed away in 1912 when he was quite young – at the age of 46. During the pre-revolutionary time, another two award-winners could have been A.S. Popov and D.I. Mendeleev (the latter could have been awarded the prize either for physics or for chemistry). But Popov died in 1905, when the radio had not yet had time to become firmly established. As for Mendeleev, he was nominated, but he passed away in 1907, and, the main thing was that, according to Nobel's will, the prize was to be given for works done quite shortly before awarding the prize, while Mendeleev's works did not meet this requirement.

---

[13] A.M. Blokh, "Sovetskii Soiuz v Interiere Nobelevskih Premii" (The Soviet Union in the context of the Nobel Prizes) (Moscow: Nauka, 2005). The translation of this book into English is going to be published by the "World Scientific" publishing company.

By the by, in 1936 the requirement to award the prizes only for 'fresh' works was abrogated.

After the revolution, physics in the USSR developed rather rapidly, but I can name only three works which should have gotten (and not merely could have gotten) the Nobel Prize in Physics, but they did not get it. Firstly, there are the works of A.A. Friedman (1922, 1924) on cosmology on the basis of the general theory of relativity. However, these works won acclaim only several years later (we can say that it happened after Hubble's work on the cosmological redshift was published, in 1929) – while A. Friedman died in 1925, when he was only 37. Maybe it is worth reminding the reader that according to the status of the Nobel Prizes, they cannot be awarded posthumously.

The Nobel Prize should have been awarded to G.S. Landsberg and L.I. Mandelshtam for their 1928 discovery of combination scattering of light. This important phenomenon is usually called the Raman effect, and it was really discovered, independently and simultaneously, by Raman. However, it was only Raman who got the prize for this discovery, in 1930. I consider this fact an unquestionable error of the Nobel Committee, but the question is not so simple – as it is also the fault of the authors (Landsberg and Mandelshtam) and of the Soviet physicists, who did not nominate them. This case is described in detail in my book "*About Science, Myself and Others*" (book II), in articles 21 and 22. Repeating myself here would be inappropriate.

In the 1930s and later (under Stalin's dictatorship) all contact with the Nobel Committees was interrupted and, in fact, banned (see the previously mentioned book by A.M. Blokh). It was only after the death of the dictator that our authorities decided to come back to, so to speak, 'the Nobel Club' and to nominee scientists from the USSR. It was decided somewhere 'above'. I only know that, perhaps about the year 1956, I was charged to prepare, along with one of my colleagues, a nomination for the prize for the discovery of the Vavilov–Cherenkov effect. Of course, it was done. We suggested the candidatures of Tamm, Frank, and Cherenkov, but submitted or supposed to be nominated, from 'above' was Cerenkov only. Such a nomination was, or would have been, patently unfair. The thing is that the leading part in the discovering and explaining of the Vavilov–Cherenkov effect[14] was played by S.I. Vavilov. However, he passed away in 1951, not even having lived to the age of 60. Thus, he could not be nominated, and the only acceptable combination was: Tamm–Frank–Cherenkov (the names are given in the order of the Russian alphabet). In book I, the article "In Memory of Ilia Mikhailovich Frank", dedicated to the memory of I.M. Frank, tells what E.L. Feinberg and I did in order to obtain a fair awarding of the prize, namely to Tamm, Frank and Cherenkov, which did take place in 1958. The details of the procedure of awarding this prize will soon (in 2009) become known.

---

[14] It is only this name, rather than the more common name 'the Cherenkov effect', that is considered fair by all the well-informed colleagues whom I know.

Later, the prize was awarded to the following scientists from the USSR and Russia: in 1962, to L.D. Landau; in 1964, to N.G. Basov and A.M. Prokhorov (half the prize); in 1978, to P.L. Kapitsa (half the prize); in 2000, to Zh.I. Alferov (half the prize, and with a foreign co-author) and, finally, in 2003, A.A. Abrikosov, A.J. Leggett and myself[15]. All who have got any part of the prize are considered to be prizewinners, and the association of the award with the country comes out of consideration of the prize-winner's citizenship. So, we can say that in all Russia has gotten 6 prizes for physics, with 9 Russian citizens becoming prizewinners (not formally including A.A. Abrikosov, since in 2003 he was already a citizen of the USA).

Of the Russian citizens who were supposed to get the prize but did not get it, I can point out only E.K. Zavoiskii. He was the first to watch, in 1942, electron paramagnetic resonance. When the authorities permitted it, E.K. Zavoiskii was repeatedly nominated for the Nobel Prize in Physics. I myself did it at least twice, if I remember correctly, after 1966, when I started to get the Nobel Committee's offers to nominate a candidate for the prize. I was going to do it further, but E.K. Zavoiskii passed away in 1976 at the age of 69. By the by, not awarding the prize to Zavoiskii cannot be considered discrimination against Soviet science, since no one else got the prize for the discovery of paramagnetic resonance either (thus, unlike the case with the combination scattering, the Nobel Committee did not commit any error).

It is possible that the Nobel Committee is making errors of a different kind than in the case of Landsberg and Mandelshtam. The thing is that often it is not only prizewinners who are the authors of the published works for which the prize has been awarded. Thus, when several authors must be crossed out in order to leave only three prizewinners, it is not easy to do it, as well as to throw out all previous works (which often exist). Of course, a detailed expert examination is carried out, but it is not an easy thing to do and it is not always objective. In such cases some authors feel that they have been passed over. I think that it is inevitable even in the case of the most conscientious approach of the Nobel Committee. In any case, before accusing the Committee of being biased or incompetent, it is necessary to have strong arguments for that. Unfortunately, they can sometimes turn up only some 50 years after awarding the prize, when all the details have been published (who nominated which person, the analysis of the works and so on).

In the year of receiving the prize, P.L. Kapitsa was 84, and I was already 87. Along with everything said above it illustrates my 'theorem' given at the beginning. I think that any discrimination against Russian physicists and, probably, representatives of other specializations, is absolutely out of the

---

[15] The prize can be given to not more than three people. However, it can be divided into two parts, which have absolutely no connection with each other. For example, P.L. Kapitsa got 'a half' of the 1978 prize for his works on low-temperature physics, and the second half of the prize was awarded for the discovery of cosmic microwave background radiation.

question. In order to get the Nobel Prizes for science, our scientists should, first of all, make good works and, of course, not to forget to nominate their compatriots, as it was, for instance, with the discovery of the combination scattering of light. "Which of the Russian physicists should be awarded the prize today?" – that is the question often asked. I do not know such a candidate, but I do know worthy candidates who could be awarded the prize (I hope that the difference between *should be* and *could be* that I mean to express is clear). By the way, I nominated one such candidate for the 2006 prize.

I hope that my giving here rather a lot of attention to the Nobel Prizes is justified by the great number of questions that have been addressed to me on this issue.

Of course, the main thing is not prizes but scientific progress and achievements. I belong to those who connect the hopes of the survival and development of civilized society with the utmost development of education and science. That is why I take to heart the fact that in Russia today science is, so to speak, out of favor or, in any case, is not given enough attention. There is no understanding of the fact that the division of science into fundamental and applied science is rather conventional, while the submission of fundamental research to the requirements of the present day and to the interests of business and officialdom is absolutely unsound. However, we must take into consideration reality, and the impossibility in Russia to spend on science colossal sums comparable to those allocated in the USA. That is why I try, as far as possible (though I now do not have much strength and possibilities remaining, as is clear from what follows), to contribute to the development of education and science in Russia. The work is carried out mostly in three directions. Firstly, the Ginzburg Foundation Physics-Uspekhi has been created, with the purpose of popularization of science (first of all, physics), mostly among young people. The point is also to get capable people interested in scientific activities, not to mention enlightenment and working from a scientific outlook. The Foundation intends, among other things, to publish a number of collections of articles and brochures aimed at achieving the indicated objectives. This has already partly been done (for all data about the Foundation and its work see www.uspekhi-fond.ru). As this is my autobiography, I will permit myself to point out that one of the things done by the Foundation was a new edition of the popular brochure "Superconductivity" jointly prepared by E.A. Andriushin and myself (Moscow: Alpha-M, 2006; the first edition of the brochure is given in [211] for the second article in this book).

Secondly, besides the activities of the Foundation, I discuss, in a number of articles and interviews, different questions connected with the development of science in the country, for example, the questions about the reorganization of the Russian Academy of Sciences, the combat against false science, (pseudoscience) and clericalism. In the latter case it is especially important to oppose the attempts of the Russian Orthodox Church to penetrate even into elementary schools. That is a manifest contradiction to the Constitution of Russia as a secular state. The corresponding material is given on the site of the journal *Physics-Uspekhi* (www.ufn.ru; hyperlink "Tribuna UFN"). I cannot but note that in the combat against false science and clericalism I work in contact with the Russian Humanist Society (RHO), which publishes the journal *Zdravy Smysl* (Common Sense), whose editor-in chief is the President of the RHO, V.A. Kuvakin. The journal is a quarterly, 40 issues have already come out, and the latest of them, No. 3, 2006 is dedicated to the 10th jubilee of the RHO and the journal. The RHO's position is that of civil (secular) humanism, and I can say that I share the philosophy and outlook of this trend (for more detail see the article by V.A. Kuvakin and this author in my book *About Science, Myself and Others* (book II, Article 35) and my interview to him in *Zdravy Smysl* No. 3, 2006).

Thirdly, I try to implement what has been said, that is, to develop a scientific trend in the field of fundamental physics, which is both very interesting and promising and, on the other hand, comparatively cheap. I mean the study of superconductivity, which is the topic of article 3 in the present book. In February 2006 I directed the corresponding considerations and material to the President of Russia, V.V. Putin.

I have not received any response yet, and most probably I will not achieve any success in that matter. Such a pessimistic estimation results not merely from my knowledge of the situation of science in our country and the fact that the sum required is not so small by our scale (it is $20 million), but, and this is the main thing, from the fact that I am ill; not only am I unable to be active enough for achieving success (not to mention that I do not know how to do it), but I am physically incapable of doing it for the reasons which are clear from what follows.

But I have gotten ahead of my story, since I have not yet written about the last stages of my life. I was undoubtedly very lucky to have been awarded the prize at the age of 87, when I still had enough energy to prepare a lecture and, as I already wrote, to go to Stockholm and take part in the corresponding events. Then, through the whole of 2004 I was quite well and in general had my usual way of life. But in fact, as I now know, a disease was stealing up on me – walking was becoming more and more difficult. And on 23 January, 2005, I found myself in hospital, still without any diagnosis made, and several days later became totally unable to walk. I was rather energetically examined, and, luckily, the renowned doctor A.I. Vorobiev quickly found out that I have a certain rare blood disease (Waldenstrom's disease). I still do not know more specifically what it is about (for some reason, I had no wish to study hematology; it is not simple, and I have become not very curious about 'someone else's' sciences). As far as I know, it is believed that if not cured, this disease can at least be stopped in its development. That has probably been achieved, but the ability to walk has not been restored. From 18 February, 2005 to 24 July, 2006 I was in hospital at the Institute of Hematology in Moscow. It is hard to lie month after month on one's back; of late they have intensively tried to restore the functions of the legs by exercises, and it is still harder. Now I am at home and can work at a table. Anyway, it is good that I manage (at least, have managed so far) to do something: give interviews over the phone and write notes concerning issues now topical for me, which I mentioned previously. For probably very few readers who know Russian, I can say that the greater part of what I have written is placed on the site www.ufn.ru, in the hyperlink "Tribuna UFN". Besides, it is impossible not to mention my work connected with the journal *Uspekhi Fizicheskikh Nauk*. English name is *Physics-Uspekhi*. I have long had connections with this journal; it started as far back as before the Second World War, and for many years I was a member of its editorial board and since 1998 I have been its editor-in-chief. In my opinion, *Physics-Uspekhi* is a good and useful journal and I do what I can for making its work successful. Of course, now that I am ill I am doing less than should be done. Fortunately, for 18 years already we have had M.S. Aksent'eva as the editorial manager and as a member of the editorial board. It is mostly due to her, along with other editorial board members, that the journal owes the maintenance of its high level. Writing about these topics and about my today's life in general in a more detailed way seems inappropriate here. In conclusion I will only permit myself to make two more remarks.

Now, my two closest friends, (mentioned in the above autobiography) have experienced, like myself, hardships connected with old age. Alas, these two friends – E.L. Feinberg and I.L. Fabelinskii – have already passed away (the obituaries were published in *Uspekhi Fizicheskikh Nauk* **176**(6) 683–684 (2006) and **174**(11) 1271–1272 (2004) respectively) [*Physics-Uspekhi* **49**(6) 659 (2006) and **47**(11) 1173 (2004)]. It pains me to think about it.

In my Nobel Lecture, which opens this book, I expressed a certain satisfaction with the fact that I had followed the way of my teachers, I.E. Tamm and L.D. Landau. Of course, I meant their way in science. Unfortunately, it turned out that at the end of my life I have also followed their way. I.E. Tamm fell ill, at the end of the 1960s, with amyotrophic lateral sclerosis and spent the last several years of his life on a breathing machine (he passed away on 12 April, 1971, when he was 75). L.D. Landau was involved in a major car crash, and after that he lived for six more years, not having the ability to work (he passed away on 1 April, 1968, at the age of 60). I am also ending my life an invalid.

Such are human fates.

21 September, 2006.

P.S. I was ninety on 4 October, 2006. In this connection, three scientific sessions embracing a wide range of topical problems were held. The scientific session of the Division of Physics Sciences of the Russian Academy of Sciences and a special meeting of the *Uspekhi Fizicheskikh Nauk* [*Physics-Uspekhi*] Editorial Board took place in Moscow, and the joint session of the

Scientific Council of the Institute of Applied Physics, RAS and the Editorial Board of the journal *Izvestia Vuzov. Radiofizika* [*Radiophysics and Quantum Electronics*] was held in Nizhny Novgorod. The contents of the reports given in these sessions were published in the journal *Physics-Uspekhi* (see the references listed at the end of Chap. 6 in this book). Furthermore, I was decorated with a Russian Federation First Class Order of Merit for the Fatherland and received a lot of greetings and congratulations on my birthday. Certainly, it is out of place here to write about this at greater length. I should like only to avail myself of the possibility of thanking all those who have congratulated me.

6

A Scientific Autobiography: An Attempt[1]

6.1 Introduction

For a professional writer, the publication of selected works or even a complete set of works in several volumes in the writer's lifetime is the norm (provided a willing publisher is found). Among scientists, however, selected works are very rarely printed during a scientist's life. The latest example that I know of is the two-volume set of Y.B. Zeldovich's selected works [1]. The main reason for this difference (this can hardly be doubted) is that typically novels, stories, and the like do not become obsolete; at any rate they remain attractive and interesting (I certainly mean only high-quality literature, not the mediocre or fly-by-night books that have often been published with enormous print runs). In contrast to this, scientific publications often overlap, they are developed and extended by later publications, or they are reflected in reviews, monographs, and textbooks. Nevertheless, a publication of collected scientific works cannot be regarded as unnecessary. (Of course, quality is the issue, but I again mean important papers, important at least at the time of their publication). There is no need to prove this when a great scientist is involved, but physicists and other scientists of lesser rank also often accomplish a great deal of very useful work in their lifetime. It would be convenient to be able to read their papers without browsing through old journals. For this reason, and also as a tribute to the author, numerous selected works are published posthumously. Assuming all this, isn't it legitimate for an author to take part in the publication of his or her own selected works while still in this world? It appears that such attempts are deemed immodest; besides, very few rise to the occasion and undertake this time-consuming job. On the other hand, a publication with the author's participation should, in general, be of much higher quality than a posthumous one. To make a long story short, I had nothing against Zeldovich

---

[1] The present paper was written in 1991 and first published in the Russian edition (Nauka, Moscow, 1992). The present English version contains a few slight changes and added references [140–169].

publishing his collected works. When he presented the two volumes to me, he said something like, 'You'll reach 70 soon too, so follow my example.' And then he added: 'This absorbed two years of my life.' No doubt, Y.B. had not devoted two years exclusively to compiling, editing, and commenting on his publications [1], but all this definitely demanded a huge amount of work. Quite a few others were also involved in this effort. I confess that Ya.B.'s advice seemed very seductive at the time. Summing things up is quite natural, once you reach a certain age. However, I soon rejected the idea of publishing my own collected works: the effort involved is enormous and hard, and it would be unlikely that anyone else could be found to help me. Ultimately, I am not sure in my case that the interest in my 'collected works' would warrant the effort. But a thought came to mind that it would be different if I wrote an article – while compiling the volume you are now reading – that would kind of give a synopsis of the nonpublished selected works and would be a variant of a scientific autobiography (I mean an account of my scientific efforts), referring to those publications of mine that I consider to be of some value. Even this approach may not be – and probably will not be – justifiable to some; however, reading it is not compulsory, and this calms me.

6.2 Classical and Quantum Electrodynamics

I got involved in theoretical physics somewhat accidentally, having come across a problem that I managed to sort out to a first approximation. The details, omitting the physics involved, are told in the memoirs about I.E. Tamm (see p. 351 in [151]). Here I shall outline only the gist of the story.

At the end of 1938, when my working life began, the atmosphere and the entire situation in theoretical physics (and in physics in general) were very different from what we see now. The number of theoretical physicists in Moscow at that time hardly reached several dozens, while today it is in hundreds. It was sufficient to drop into the institute library once a week to see all the latest publications – several relatively thin issues of physics journals in German and English. Now, even in Germany, all physics journals publish papers almost exclusively in English, which has become a sort of neo-Latin (in its role of the international language of science). The total number of journals and the number of pages in a journal have increased by a factor of several tens. Half a century ago it was possible to follow the whole of physics, while today this is just not possible. Nevertheless, browsing through all new journals was an obsession of mine, and I 'kept afloat' for quite a while; however, I had to give up recently – the amount of time available and my stamina are in decline while the number of journals keeps rising.

Returning to my own work, I should say that it started not with reading journals but with an attempt to implement an 'idea' that was stimulated by certain experiments with canal rays (*Kanalenstrahlen*) that I tried to reproduce for my diploma project. The experiments themselves are of no interest now, so it is sufficient only to formulate the essential part of the problem. Consider an atom that emits spontaneously at a frequency ω_0 with the same probability in the directions z and $-z$. Assume now that a charge (an electron or an ion) moving at a velocity v along the z axis impinges on the atom. We shall expand the electromagnetic field of the charge in plane waves of the type $A \exp[i(\omega t - \boldsymbol{k} \cdot \boldsymbol{r})]$. It is easily shown that in this case $\omega = \boldsymbol{k} \cdot \boldsymbol{v} = k_z v$. At the time I already knew of the existence of stimulated emission: a photon impinging on the atom at a frequency equal to the transition frequency ω_0 increases the probability of emission in the direction of the impinging photon. As a result, the probability of radiation in the direction of $\boldsymbol{v}$ appears to be higher than in the direction of $-\boldsymbol{v}$. This conclusion is wrong, since the field of a moving charge is not equivalent to an ensemble of photons in a vacuum, for which $\omega = ck$ (c is the velocity of light); as I mentioned, for the field of the charge we have $\omega = k_z v \le kv < kc$. Even now, the textbooks familiar to me use words that identify, in quantization, the transverse electromagnetic field with the photon field. This is patently wrong, of course: the electromagnetic field of a moving charge, carried along by it, is not the same as an ensemble of photons, or, in strict physics terms, the carried field is the field of off-mass-shell photons, that is, the photons which do not satisfy the condition $\omega = ck$. The results of quantum-mechanical calculations are nevertheless correct because 'mathematics is cleverer than man' (for a more elaborate treatment of this, see Chap. 1 of [2]). Even though I often planned to tackle the problem of the quantization of the carried field, I was never able to – and this possibly was not accidental: quantum-electrodynamics formalism (or any other formalism) is not my pasture. It is nonetheless strange that this problem – a simple one in principle – has not been properly elucidated anywhere.

The attempt to explain the spatial asymmetry of the emission intensity for charges flying past excited atoms was thus a failure. However, when studying quantum electrodynamics, I found in a paper by V.A. Fock [3] and in a paper that extended his work [4] a statement that was a revelation to me: a uniformly moving charge emits radiation. In classical electrodynamics we are used to thinking that a charge moving at constant speed (in vacuum) cannot emit. My first scientific result deals with explaining this seeming contradiction [5]. The contradiction stems from the differently posed problems: in classical electrodynamics we normally speak of a stationary problem – a charge always moves at a constant velocity v and indeed does not emit. But in quantum electrodynamics (at that time it was more often called the quantum theory of radiation [6]) the formulation of the problem, using perturbation theory, was as follows: there is a uniformly moving charge (electron) at time $t = 0$, but no photons. Photons then appear at $t > 0$ and the charge emits radiation. In fact, the situation would be the same in the classical case if we assumed that the transverse electromagnetic field was totally absent at $t = 0$ and the charge was in motion. Physically, this means that the charge accelerates abruptly (instantly) at the moment $t = 0$ and gains a velocity v. This means, however, that the charge should emit both the field it carries and a certain radiation field due to the acceleration of the charge. As I was able to show in [5], the classical and quantum calculations coincide in this simple situation. The technique I used was the so-called Hamiltonian method that I learnt from Heitler's book [6]; I am still in love with it – it is simple and visually clear. The main part of my first publication [5] can be found in Chap. 1 of an easily accessible book [2] (to which I refer the reader). Inspired by the fact that I could work with and even clarify something in quantum electrodynamics, which at that time was regarded as physics's 'frontier', I applied the same Hamiltonian method to a study of quantum-electrodynamic divergences [7]. Then I came across a statement in the literature which assigned spontaneous emission to the action of zero-point fluctuations in the vacuum. This would mean, however, that spontaneous emission is a purely quantum phenomenon, which is obviously wrong. My third paper [8] discussed this problem and was published in the same year (1939). This not excessively complicated matter (the nature of spontaneous emission) is still being discussed. I also returned to it in a methodological note [9].

This initial stage of my research also includes a paper [10] that today can only raise people's eyebrows. The thing is that Heitler in his book [6] (in its first edition), and all other sources known to me at the time, operated with the Lorentz gauge of the electromagnetic potential div $\boldsymbol{A} + (1/c)(\partial\phi/\partial t) = 0$. The longitudinal field could then be singled out only by a special transformation. I was able to show that the result could be immediately obtained by using the Coulomb gauge div $\boldsymbol{A} = 0$. Nowadays any student knows this, but in 1939 even I.E. Tamm and V.A. Fock, both world-famous theoretical physicists, were not aware of this result and recommended that I publish it [10]. Several years later I discovered that the Coulomb gauge had been successfully applied before me. This is, least of all, an attempt to establish priorities; I describe it more to illustrate the state of the theory in 1939.

6.3 Radiation by Uniformly Moving Sources (the Vavilov–Cherenkov and Doppler Effects, Transition Radiation, and Related Phenomena)

The study of uniformly moving sources can be considered a special chapter of electrodynamics even though it cannot be reduced to electrodynamics only (analogues are known in acoustics and the theory of any field). Whatever the reason, I am still in love with this range of problems more than with any other (love is not a word that is used often in science, but I regard this as no more than a tradition or convention). The explanation possibly lies in the fact that the theory of the Vavilov–Cherenkov emission of radiation was constructed in 1937 by I.E. Tamm and I.M. Frank [11] before my very eyes; to be honest, I was very sorry for being slightly 'late' and thus not taking part in thinking over the nature of the Vavilov–Cherenkov effect, discovered in 1934. Anyway, I never forgot the Vavilov–Cherenkov effect and proposed, in the paper already mentioned previously [8], another method of arriving at the Tamm–Frank result. In their paper [11], Tamm and Frank calculated the field of a charge moving uniformly in a medium and then found the flux of the Poynting vector across a cylindrical surface surrounding the trajectory of the charge. In my approach, I used the Hamiltonian method and calculated the emitted energy, which is simpler. The condition for the emission of Vavilov–Cherenkov radiation is obtained immediately, since the equations for the amplitudes of the field oscillators have the form

$$\ddot{q}_\lambda + \omega_\lambda^2 q_\lambda = \sqrt{4\pi}\,\frac{e}{n}\,\boldsymbol{e}_\lambda\!\cdot\!\boldsymbol{v}\exp\left(-i\boldsymbol{k}_\lambda\!\cdot\!\boldsymbol{r}_i\right), \qquad (6.1)$$

where $\omega_\lambda^2 = (c^2/n^2)k_\lambda^2$, $\boldsymbol{e}_\lambda$ is the polarization vector of the radiation, $\boldsymbol{v}$ is the velocity of the charge at a point $\boldsymbol{r}_i(t)$, and n is the refractive index of the medium (for details of what I describe here and below, see Chap. 6 of [2]).

In terms of the Hamiltonian method, a charge (or any other source) emits if the amplitudes q_λ increase in time, and this happens for large times t at the resonance condition, that is, when the right-hand side of (6.1) contains the frequency ω_λ. For a uniformly moving source we have $\boldsymbol{r}_i(t) = \boldsymbol{v}t$ and the condition for radiation becomes $\omega_\lambda = ck_\lambda/n = \boldsymbol{k}_\lambda\!\cdot\!\boldsymbol{v}$, or

$$\cos\theta = c/(nv)\,, \qquad (6.2)$$

where θ is the angle between $\boldsymbol{k}_\lambda$ and $\boldsymbol{v}$. However, the condition (6.2) is precisely the Vavilov–Cherenkov condition. Obviously, $n = 1$ in vacuum, and Vavilov–Cherenkov-type emission at $v < c$ is impossible. However, if the amplitude q_λ and its derivative $\dot{q}_\lambda \equiv dq_\lambda/dt$ equal zero at $t = 0$, then at later times $q_\lambda(t)$ and $\dot{q}_\lambda(t)$ are nonzero; for the adiabatic switching on of the interaction or the slow acceleration of the charge, this corresponds to the 'emission' (formation) of the carried field of the charge, as I already discussed in Sect. 6.1.

Calculating $q_\lambda(t)$ and $p_\lambda = \dot{q}_\lambda(t)$ and then the field energy

$$H_{\text{tr}} = \int \frac{\epsilon E_{\text{tr}}^2 + H^2}{8\pi}\,dV = \sum(p_\lambda\dot{p}_\lambda + \omega_\lambda^2 q_\lambda\dot{q}_\lambda)\,, \qquad (6.3)$$

we obtain an expression for the radiation intensity or, specifically for uniform motion, for the Vavilov–Cherenkov radiation. Naturally, the result of this calculation (see [2,8]) coincides with that obtained in [11]. The third method of calculation is to find the work done by the force of radiation friction (i.e., by the force $e\boldsymbol{v}\!\cdot\!\boldsymbol{E}(\boldsymbol{r}_i)$, where $\boldsymbol{E}(\boldsymbol{r}_i)$ is the field acting on the charge; see Chap. 14 of [12]). Although the three outlined methods of calculating the emission intensity yield the same result for the Vavilov–Cherenkov effect, they are in reality far from identical. In the general case, it is natural that the energy flux across a surface, the change in the energy of the field in a volume, and the work of the radiative friction (reaction) force are not equal to one another (for details, see Chap. 3 of [2]).

Which of these methods is more convenient and efficient depends on the type of problem and on the quantity to be found. To be specific, I shall choose emission of radiation in an anisotropic medium, for example, in a crystal. The equations of electrodynamics in an anisotropic medium were definitely well-known already fifty years ago. Nevertheless, they were applied only to describe the propagation of 'free' electromagnetic waves (including light waves) – this constitutes the contents of, say, crystal optics. But how does a dipole (an oscillator) emit in a crystal? I was unable to find an answer to this simple question in the literature (I still do not know if this problem had been solved by anybody before my paper of 1940 [13]). We are dealing here with a generalization to the anisotropic case of the well-known (for a vacuum or an isotropic medium) expression for the energy emitted per unit time into a solid angle $d\Omega$:

$$\frac{dH_{tr}}{dt} = \frac{e^2\omega_0^4 a_0^2 n}{8\pi c^3} \sin^2\theta \, d\Omega \,, \tag{6.4}$$

where a_0 is the amplitude of small oscillations of the charge e at a frequency ω_0, and θ is the angle between the axis of the dipole and the direction of observation. The easiest way to derive (6.4) is to apply the Hamiltonian method (essentially by expanding the field in plane waves), but it is usually obtained from the general solution of the field equations by using retarded potential (see, e.g., Sect. 67 of [14]). In an anisotropic medium, it is quite a challenge to get up and write down a solution for the potentials, while the Hamiltonian method allows obvious generalization. One needs to expand now into 'normal' electromagnetic waves that can propagate through the anisotropic medium under consideration. The equation for the amplitudes of these waves is similar to (6.1). Subsequent calculations are also simple and yield the following result (see Sect. 6 of [2], and [13]):

$$\frac{dH_{tr,l}}{dt} = \frac{e^2\omega_0^4(\boldsymbol{a}_l\cdot\boldsymbol{a}_0)^2 n_l^3}{8\pi c^3} \, d\Omega, \tag{6.5}$$

where $\boldsymbol{a}_0$ is the charge oscillation amplitude, $\boldsymbol{a}_l$ is the appropriately normalized polarization vector of the normal wave l, and n_l is the refractive index corresponding to this wave (of course, (6.5) transforms to (6.4) in an isotropic medium). The problem of Vavilov–Cherenkov emission in crystals [15] is solved by the same method (there was an integration error in [15]; see [16]).

In relation to the theory of the Vavilov–Cherenkov effect, I shall also mention some problems of emission in channels and slits and also emission by various dipoles (magnetic, electric, toroidal). I do not command enough space here to describe this, but I can refer the reader to reviews [2,17] with relevant bibliographical references.

In view of all this, it is quite natural that already at the first stage of my work (in 1940) I had succeeded in constructing the quantum theory of the Vavilov–Cherenkov effect [18]. If one uses the concept of 'photons in the medium' with energy $\hbar\omega$ and momentum $(\hbar\omega n/c)\boldsymbol{k}/k$ (this occurs automatically in the quantization of the electromagnetic field in a medium), the energy and momentum conservation laws yield the expression

$$\cos\theta_0 = \frac{c}{n(\omega)v_0}\left[1 + \frac{\hbar\omega(n^2-1)}{2mc^2}\sqrt{1 - \frac{v_0^2}{c^2}}\right],\qquad(6.6)$$

where θ_0 is the angle at which the particle, which moved uniformly at a velocity v_0 (before emitting a photon), emits a photon with energy $\hbar\omega$. As could be expected, the condition of emission (6.6) transforms to the classical condition (6.2) if $\hbar\omega/(mc^2) \ll 1$. In optics, $\hbar\omega/(mc^2) \lesssim 10^{-5}$ even for electrons, and therefore the quantum approach to Vavilov–Cherenkov emission is of no practical importance. L.D. Landau immediately pointed this out (I describe this on p. 382 in [151]). Nevertheless, the quantum treatment of emission in a medium proved to be productive. The point is that conservation laws not only make it possible to find the relation between θ_0 and ω, but also yield the direction of transition between energy levels in the case of an atom, etc. (for example, we may find for levels 1 and 2 with energies E_1 and E_2 that the transition associated with radiation occurs from level 1 to level 2, not the other way around). In view of this, it is immediately established [19] that in the range of the anomalous Doppler effect (see also [2, 17]), emission entails excitation of the emitter. This remark is quite important for the interpretation of the anomalous Doppler effect and also for understanding the nature of the excitation of an accelerated 'detector' [20].

In addition to the Vavilov–Cherenkov and Doppler effects, uniform motion of the source may produce transition radiation, which I.M. Frank and myself [21] discussed in 1944. In this case, the velocity of the source v may be below the phase velocity of light $v_{\text{ph}} = c/n$ but the inhomogeneity of the medium along the trajectory of the emitter (charge, etc.) becomes an important factor. The best approach is to relate the transition radiation to the variability of the parameter vn/c, even though this is rather formal: $n = 1$ in a vacuum, and radiation is emitted only in response to acceleration, that is, when v/c changes; in a medium, however, emission also occurs at constant v but varying vn/c. It was found that transition radiation is an effect 'rich' in consequences (if an effect can be called 'rich') and manifests itself in a number of forms (spatial and temporal nonuniformity, transition scattering, transition bremsstrahlung); it plays an important role in plasma physics, for designing special (transition radiation) counters, etc.

I explained at the very beginning of this section the reason why I devoted so much space in this section to radiation by uniformly moving sources, out of proportion to its importance. Enough of this, though; I cannot go into details of transition radiation and related phenomena and shall have to cite the reviews [2, 17, 22, 140], where references to the original papers can be found.

6.4 About This Chapter

No one knows the best plan and theme to which a scientific autobiography should be written (or at least which is recommended). It would be preferable not to reduce it to a mere list of problems and publications; one wishes an internal logic behind the author's efforts. This is obviously easier if one subject dominates all work (Chap. 2 in this book may serve as an example). The material can then be unfolded in a time-ordered sequence. As for me, I worked in very many areas and, by the way, regard this possibility as one of the most attractive features of theoretical physics. The reasons causing a switch from one problem to another varied: sometimes a certain logic of moving on, sometimes accidental stimuli, or the pressure of such a powerful factor as the war going on, or other 'extraneous' factors.

The preceding pages clarify how I got into theoretical physics. Even if it was an accident, I started with a problem that at that moment was at the hub of physics – quantum electrodynamics, and elementary-particle physics. Fortunately, other things interested me as well and my eyes were open. Hence, having discovered the effectiveness of the Hamiltonian method in the vacuum [5, 7], I immediately applied it to the electrodynamics of continuous media [8, 13, 15, 18] and thus 'locked on to' the theory of emission by uniformly moving charges. The first result in this field was obtained in 1939 [8] and the last (at this moment – but very likely this qualification is irrelevant) dates back to 1985 [23]. But, realistically, the theory of the Vavilov–Cherenkov effect [8, 15, 17, 18] was nevertheless a sideline with me. My main field from 1940 on was the theory of particles with higher spins (I shall discuss this in Sect. 6.5). However, the war that for us started on June 22, 1941 pushed us into looking for more practical and immediately useful applications of our skills. I remember how we, the theoreticians of the P.N. Lebedev Physical Institute (FIAN) of the Academy of Sciences of the USSR, were questioning everyone, 'What useful work could we do for the defense of the country?' – this was not at all obvious at the time, and the transition to military orientation was not prepared. For instance, I.E. Tamm started calculations that were needed for the demagnetization of ship hulls (to counteract the threat of magnetic mines), and B.A. Vvedensky gave me advice on analyzing the spreading of radio pulses reflected from the ionosphere. My first 'defense-oriented' paper was devoted to this very topic [24]. Another piece of applied research was the theory, developed together with I.E. Tamm, of electromagnetic processes in layered magnetic cores (the aim was processes in antennas) [25]. Judging by the list of my publications in the war years (this list can be found in [26]) – and this is supported by what I can recall – I was aware of the rather dubious practical value of my work on propagation of radio waves, and thus did not stop working on the relativistic theory of particles with higher spins or in some other areas. Nevertheless, the theory of wave propagation in plasmas (e.g., in the ionosphere) became an important part of my work – and my life – for many years to come.

If I went in for more details, this scientific autobiography would transform to an ordinary one, something I would not like to happen![2] I shall only

---

[2] This was to some extent made in the paper "Notes of an Amateur Astrophysicist" written by invitation of the journal *Annual Review of Astronomy and Astrophysics*

mention that the work on wave propagation in plasmas led to an interest and work in radio astronomy and then in certain areas of astrophysics, including the astrophysics of cosmic rays and gamma astronomy. I shall describe these directions of research later in the article.

Another line was stimulated by L.D. Landau's theory of superfluidity [28]. In 1940 (or thereabouts) I listened to Landau's talk on this subject and may have learned for the first time that the nature of superconductivity was not yet clear. It was only natural to try and do something in this field. From that time on (in fact, the first publication appeared in 1944) until now I have been involved with the theories of superfluidity and superconductivity. In the intervals between my electrodynamics, spin, plasma, and superconductivity studies, there surfaced astrophysics, ferroelectricity, crystal optics with spatial dispersion taken into account, and some other topics.

All in all, the path was tortuous, with the causes of the twists listed previously and partially clarified. This is the reason why this article is somewhat fragmentary.

6.5 Higher Spins

Even now – let alone half a century ago – physical theory is dominated by the consideration of particles with spins 0 and 1/2. True, we need to add to this the photon, a particle with spin 1 but with zero mass. At the same time, there never was and still isn't a basis for denying that particles may exist with higher spins (3/2, 2, etc.) or with spin 1 but non-zero mass (in fact, the $W^{\pm}$ and Z^0 bosons discovered in 1983 are particles of this very type, and it is difficult to doubt the existence of the graviton – the quantum of the gravitational field, with spin 2 and zero mass). The relativistic equation (Dirac's equation) for spin-1/2 particles was discovered in 1928 and that for spin-0 particles even before that, in 1926. It is thus natural that theoreticians started to study equations for particles with the higher spins 1, 3/2, 2, etc. (and arbitrary mass) interacting with the electromagnetic and other fields, already in the 1930s. A significant difference was found between the behavior of these equations and those written for the spins 0 and 1/2. In these latter cases, divergent expressions appeared in the higher approximations of perturbation theory (the renormalization method was developed only in 1948), but quite reasonable

---

[27] (the Russian version of the paper was published in [153]; the present book includes my autobiography which was written in connection with awarding the Nobel Prize) (Chap. 5 in this book). I shall only note that it is in connection with the study of wave propagation that since 1945 I had for several years been head of the chair (as a part-time worker) of radio wave propagation at the Radio Faculty of Gorky State University (GSU) (now Nizhnii Novgorod), where I had a number of post-graduate students. It is in Gorky that, together with those post-graduate students and co-workers, we wrote many papers on wave propagation in plasma, radio astronomy and some other subjects.

results are obtained in the first nonvanishing approximations of perturbation theory, for example, for scattering of light by spin-0 or spin-1/2 particles. In contrast to this, even the first-approximation results for higher-spin particles yielded blatantly incorrect expressions, for instance, the unlimited growth of cross sections with increasing energy. The analysis of these difficulties (known at the time as 'difficulties of the second kind') was at the center of attention at the end of the 1930s. I also tackled this problem in 1940 and came to a conclusion that the unlimited growth of cross sections (say, for the scattering of light by a spin-1 particle) is due to the fact that the reaction of the particle's own field on the motion of its magnetic moment is insufficiently taken into account [29]. The classical nonrelativistic analysis leads then to a conclusion that taking the reaction of the particle's own field on the magnetic moment into account is in some sense equivalent to considering the equation for a top that can be found in any spin state [29, 30]. Briefly speaking, a hypothesis was born (not only to me; see references in [30]) that in order to eliminate 'second-kind obstacles' one has to consider in the equations the excited spin states of particles, that is, the analysis should not be limited to equations with a single spin state. This conclusion was of heuristic value; in order to build up a theory it was necessary to construct relativistic equations for particles that can be in various spin states. This was the problem I chose to work on, with special emphasis on $(1/2$–$3/2)$ particles that could have a spin of either $1/2$ or $3/2$ [30]. The equation that I constructed and analyzed for the $(1/2$–$3/2)$ particle coincided exactly with the one obtained by Bhabha several years later [31] (see also [34]). The relativistic theory of particles with multiple spin states is of a certain interest, but it would be natural to consider all spin states simultaneously. I.E. Tamm and myself chose precisely this approach but, to be specific, tried to work out a relativistic theory of a top, or rather to work on a certain analog of such a theory. We spent a lot of energy on this work but decided to publish it only in 1947 [32], because we could not arrive at any results valuable for the physical picture. We started with the equation

$$\left(\Box - k^2 + \frac{\beta}{2} M_{ik} M_{ik} \right) \Psi(x_i, u_i) = 0 \,, \tag{6.7}$$

where

$$M_{ik} = u_k \frac{\partial}{\partial u_i} - u_i \frac{\partial}{\partial u_k} \,, \quad u_i u_i = r^2 \,, \quad \Box = \frac{\partial}{\partial x_i} \frac{\partial}{\partial x_i} \,,$$

for a function Ψ that depends on the ordinary coordinates (the four-dimensional vector x_i) and the new four-vector u_i – the internal degrees of freedom of a particle. The mass spectrum of (6.7) was found to be infinitely degenerate and its solutions to be transformed as infinite-dimensional representations of the Lorentz group. To remove the degeneracy, we tried to impose another equation on Ψ; we also considered some other equations (see [32–34] and references therein).

I do believe that this direction was not devoid of interest and was methodically valuable too, since we treated the case of all spin states, internal degrees of freedom (in fact, we dealt with a nonpoint-like particle, since x_i can be regarded as the 'center of mass' of two points separated by a 'distance' u_i), and infinite-dimensional representations. As far as I can see, these aspects still attract the attention of some.

I devoted a lot of time and effort to studying relativistic wave equations (see [30, 32, 34]), and they were not wasted (at least I don't think they were), even though the work was not completed. At the same time, I feel happy that I realized (at the right time or at least with not too long a delay) the advisability of leaving this field: it demanded greater mathematical powers than I possessed.

6.6 The Propagation of Electromagnetic Waves in Plasmas (in the Ionosphere). Radio Astronomy

I have mentioned that in 1941 I started working on wave propagation through the ionosphere, and that the first problem I chose was the evolution of the pulse shape when waves are reflected from an ionized layer [24]. What followed was, I may say, a systematic assault on the entire range of related topics. I shall select the following results: proving that the acting field E_{ef} in a rarefied plasma (and also in the ionosphere) equals the mean macroscopic electric field E; finding the effect of the magnetic field of the Earth on wave reflection from the ionosphere; and an analysis of wave absorption, weakly nonlinear effects, etc. As in many other cases, I tried to present the entire body of material in a systematic manner. This was useful for teaching the courses at Gorky State University, and led to publishing a monograph [35] and then a less specialized book [36] with more than 1200 bibliographic references. The reader can find the required papers and results cited there. I'd rather not describe the results in any detail now; I shall only mention the effect of signal 'tripling' occurring at small angles between the magnetic field and the gradient of the electron concentration (this means small angles with the vertical direction in the case of the ionosphere; see Sect. 28 of [36], and [37]). I also regard as valuable the analysis (in a quasi-hydrodynamic approximation) of the effect of ions on wave propagation in plasmas over the entire frequency range. This allows us to understand the specifics of the high-frequency case and, which is even more important, the type of transition to the magnetohydrodynamic approximation [38] (unfortunately, this paper was written in the period when we could publish in the Russian language only, and translations into English had not yet been started abroad; as a result, [38] went almost unnoticed and was later 'overlapped' by publications in the West). Finally, I would like to mention something that was not related to the ionosphere: I succeeded in taking into account the effect of a magnetic field on wave propagation in the atmosphere due to the magnetic moment of oxygen molecules [39]. It was found that the decisive role here is played by the stimulated emission of radiation – a fact that was far from obvious at the time.

I need to point out that in the first approximation a plasma under ionospheric conditions can be regarded as 'cold', that is, that the effects of spatial dispersion can be ignored. For this reason, in [36] I considered mostly (but not exclusively) a 'cold' plasma. A more detailed presentation of the theory of wave propagation in 'hot' plasmas was given in [40] but by that time I had almost left plasma research and the monograph [40] was mostly written by A.A. Rukhadze.[3] Let me mention an article [154], devoted to the critical notes about a history of plasma research in USSR.

My plasma studies proved useful during my relatively short work on the theory of controlled nuclear-fusion reactors. Certain clarifications are needed here. In 1947 I.V. Kurchatov involved I.E. Tamm to work on the nuclear-fusion problem (that is, the study of the possibility of creating a hydrogen bomb). I was at the time Tamm's deputy in the Theory Department and quite naturally started working on the problem, as did some other colleagues at the department, including A.D. Sakharov. At the beginning our work, even though treated as highly classified, remained quite abstract in nature. Soon, however, two ideas were born – one to me and the other to Sakharov – and this changed the situation drastically. Forty years have passed since that day but the idiocy of our life is such that this work is still regarded as classified![4] I shall have therefore to limit my story to a remark, that in 1950 Tamm and Sakharov left to work in pretty remote places while I, as a security risk,[5] stayed in Moscow at the head of a small 'support group,' still with a sentry at the door. The only interesting topic that I worked on at that time 'along the classified lines' was precisely an analysis of some aspects of controlled nuclear-fusion reactors. In 1952 (or maybe at the end of 1951, my memory

---

[3] We wrote this book (article) by request of the "Handbuch der Physik", and I did not want to decline such an invitation.

[4] After A.D. Sakharov's death, the authorities dared to declassify some things, and the magazine *Priroda* published in 1990 some articles by V.I. Ritus and Y.A. Romanov (No. 8, p. 10 and p. 20) where the history of the development of the hydrogen bomb is outlined. These papers mention that I suggested using ^{6}Li in the bomb. Owing to the reaction ^{6}Li $+ n \rightarrow t +^4$ He $+ 4.6$ MeV that is referred to on p. 146 of this volume, radioactive tritium can be regenerated. As far as I am aware, the use of ^{6}Li in hydrogen bombs is regarded as very important in the literature.

[5] My future wife was accused of counterrevolutionary activities and arrested in 1944 (before we had met) but, having spent a year in jail and a labor camp, was amnestied and exiled to the town of Gorky (rather, her official residence was set in a nearby settlement). I met her in Gorky when I started to teach at Gorky State University, and we married in 1946. All my attempts to obtain permission for my wife to move to Moscow failed, and she was able to return to Moscow only after the next amnesty, in 1953; she was declared innocent ('rehabilitated') in 1956 (for details, see [27]. By the way, I believe that only my participation in the hydrogen bomb project has saved my life, or at least prevented an arrest.

is not too firm on this), someone decided that controlled fusion was such a supersecret thing, and that I was such a risk that I was not allowed to read my own research reports. Fortunately, Stalin's dictatorship came to an end soon – on March 5, 1953 – and no order to exclude me from science followed (as far as I can judge, this could have been very probable at the time). However, I never returned to nuclear-fusion research, even after this work was partly declassified in 1956 on I.V. Kurchatov's initiative. In 1962, Iwas able to publish my old reports in the area of nuclear fusion [41].

The number of interrelations and links in science (and not only in science) is very large. One thing stimulates another. I can illustrate this rather trivial remark by mentioning how I went into astronomy. N.D. Papaleksi, who thought of the radiolocation of the Sun, asked me at the end of 1945 or the beginning of 1946 to calculate the conditions for the reflection of radio waves from the solar atmosphere. This suggestion was made to me for the obvious reason: the solar atmosphere, and its corona as well, is a huge ionosphere, and I had all the necessary formulas ready. Calculations showed that radiolocation of the Sun would be very difficult since radio waves should be strongly absorbed before they reached the reflection 'point' (I did not consider reflection by inhomogeneities, and assumed the surface at which the refractive index was $n = \sqrt{1 - \omega_p^2/\omega^2} = 0$ as the level of reflection). This immediately implied a more interesting conclusion, which I indeed presented in [42] (for details, see [36]). Namely, the source of solar radio emission is not the photosphere but the corona or – for shorter waves – the chromosphere. A hypothesis was known at the time that the corona is quite hot, maybe as high as 1 million degrees was known at the time (we know now that the temperature of the photosphere is around 6000 K). The temperature of the solar radio emission from the corona (at wavelengths of 1 m or longer) was thus predicted to be very high, even in equilibrium conditions.

Cosmic radio emission was first detected in 1931–1933 (the first paper was published in 1932). However, only a few papers were devoted to radio astronomy before the end of World War II (until 1945–1946), and its significance and potential were greatly underestimated. Other authors also came to the conclusion of a high temperature of solar radio emission at about the same time, in 1946, and, more importantly, this was confirmed by observations (see references in [27, 43]). What happened was virtually a radio astronomical explosion, mostly due to the transition to peacetime and the progress in radio technologies during the war years.

Nowadays it may be difficult to believe that the angular resolution of radio telescopes at that time did not even reach 10 angular minutes. N.D. Papaleksi therefore suggested a study of the radio emission from the corona during the total solar eclipse of May 20, 1947, using the Moon as a 'shield' to help resolve different regions of the solar atmosphere. An expedition to Brazil, organized with this in mind, did solve the problem and confirmed the solar origin of the 1-m band radio emission from the Sun (see the papers cited in [43]). I was a participant in the 1947 expedition to Brazil and paid a great deal of attention to radio astronomy. Furthermore, radio astronomical research proved to be a favorite in Gorky, where we worked a great deal on it (especially together with V.V. Zheleznyakov). As in the case of ionospheric research, I shall not go into the details of what we achieved (see the relevant references in [27,36,43]). In addition to solar radio emission (aspects of propagation and generation of radio waves), I shall mention the proposal to use the diffraction of radio waves at the lunar edge to improve angular resolution, and an analysis of the causes of ionospheric and exoatmospheric flickering of cosmic radio emission. The lack of astronomical education (in plain words, my astronomical illiteracy; see [27,44] and also p. 285 in [151] for details) stopped me from any serious work in nonsolar radio astronomy until 1950, when the synchrotron emission hypothesis was published in the literature [45]; this hypothesis connected the nonthermal cosmic radio emission to synchrotron radiation by relativistic electrons. The synchrotron mechanism of radiation was totally unknown to astronomers, and even seemed to be a speculative proposal. Consequently, the nonthermal cosmic radio emission was interpreted for a rather long time in terms of the activity of hypothetical stars active in the radio frequency band. As for myself, I immediately formed a high opinion of the synchrotron hypothesis and began to expand and advertise it [46]. The impossibility of taking part in international conferences in those years thwarted a speedy clarification of the situation. For instance, my paper on cosmic synchrotron radio emission sent to the Manchester symposium on radio astronomy (1955) was not even published. However, the synchrotron mechanism was a recognized one at the Paris symposium in 1958 (I was again unable to attend), and even my 'virtual' talk was published (see references in [27,43]). It is quite likely that the delay in the West in understanding the role of synchrotron radio emission even brought physicists and astrophysicists in the USSR some benefit, in the sense that it was possible in those difficult times to obtain, without hard competition, and publish a number of results on both radio astronomy and the origin of cosmic rays.

6.7 Cosmic-Ray Astrophysics. Gamma-Ray Astronomy. Selected Astrophysical Results

Cosmic rays were discovered in 1912 (in fact, this date is approximate). Subsequently, cosmic rays were studied for many years mostly in their nuclear physics aspect, that is, because of the presence of high-energy particles. The astrophysical aspect, or, to be precise, the aspect of the origin of cosmic rays, was overshadowed. The main cause of this situation can be found in the fact that the primary cosmic rays could be studied only close to the Earth, or, rather, high in the stratosphere. In view of the high degree of isotropy of cosmic rays (the effect of the terrestrial magnetic field can be taken into account), nothing can be inferred about their sources. The discovery of the synchrotron nature of the main part of the nonthermal cosmic radio emission made it possible to relate the radio astronomical data to the electron component of cosmic rays far from the Earth. It became clear that cosmic rays exist both in our Galaxy and in other galaxies, for instance, in the shells of supernovae. This is how the astrophysics of cosmic rays was born [47], and this is how work in radio astronomy led me, beginning with the publication of [46], to the astrophysics of cosmic rays. The results obtained were presented in detail in the monographs [48, 49] (see also my talk "Astrophysical Aspects of Cosmic-Ray Research (the First 75 Years and Prospects for the Future)" [50], where references to numerous papers are given, as they were in [47]). For this reason (and owing to the fact that this chapter has grown too large) I shall not go into discussing the details of the problems that were discussed and are still discussed now (see [142], the article on p. 457 in [151], and Chap. 1 of this book). I only need to remark that the definition of the term *cosmic rays* nowadays includes only high-energy charged particles of cosmic origin (say, those with kinetic energy E_k above 100 MeV). With this definition, cosmic-ray astrophysics does not cover such important new areas of astronomy as gamma astronomy and astronomy of high-energy neutrinos.[6] At the same time, all these aspects are tightly interconnected (this is also true to a certain extent for X-ray astronomy and for the optical and radio emission of synchrotron origin). Gamma astronomy is bound especially closely to cosmic-ray astrophysics (referred to more often in the English-language literature as the 'origin of cosmic rays'). The thing is, we can extract only information on the electron component of cosmic rays directly from the radio astronomical data (because the cosmic radio emission is produced almost solely by relativistic electrons and positrons); at the same time, electrons make up only about 1% of the cosmic rays (which mostly consist of protons and heavier nuclei). In reality, however, paying the price of certain assumptions about the electron component, one can get to the proton–nuclear component of cosmic rays. At the same time, the study of cosmic gamma radiation (i.e., the use of gamma astronomy techniques) yields direct information on the proton–nuclear component of cosmic rays far from the Earth (this mostly means the gamma radiation emitted in the decay of π^0 mesons born in collisions of cosmic rays with nuclei of the interstellar medium). It is natural that we (by this, I mean myself and a number of my coauthors) work simultaneously in cosmic-ray astrophysics and gamma astronomy [2, 48–51, 135].

Astronomy has been transformed before the very eyes of the people of my generation: it was once exclusively optical and now covers all frequency bands; add to this cosmic-ray astrophysics and, in the foreseeable future, the astrophysics of high-energy neutrinos. I was lucky in that I started doing, along with physics, the 'new astronomy' quite early (in 1946).

---

[6] These fields of astrophysics, together with cosmic-rays astrophysics, are sometimes called high-energy astrophysics.

Having established myself in astronomy, I could not limit my curiosity to radio astronomy and high-energy astrophysics only, but looked into some other fields as well. I shall mention forays into the collapse of a magnetic star 4*, ways of verifying general relativity, heating of intergalactic gases, and the superfluidity of neutron stars (see references in [26, 27]).

6.8 Scattering of Light. Crystal Optics with Spatial Dispersion Taken into Account

Scattering of light was the central topic at the chair and laboratory headed by G.S. Landsberg at Moscow State University. I was a student in that laboratory; later, I gravitated, let us say, to those physicists who grouped around L.I. Mandelshtam (N.D. Papaleksi, G.S. Landsberg, I.E. Tamm, A.A. Andronov, and others).[7] The light-scattering problem was therefore quite familiar and close to me. As a result, I wrote several papers on the subject: on light scattering in Helium II [52] and in 'ordinary' liquids [53] and, at last, scattering close to second-order phase transition points (applied first of all to the $\alpha \leftrightarrows \beta$ transition in quartz). It was assumed in the past that this $\alpha \leftrightarrows \beta$ transition was a second-order phase transition or a first-order phase transition that was very close to a second-order phase transition, in other words, that it was very close to the tricritical point. In fact, as we know now, a new, nonuniform phase appears in quartz in the narrow temperature interval close to the $\alpha \leftrightarrows \beta$ transition. On the whole, the pattern of phase transitions near the tricritical point in a solid is fairly complex and multifaceted. This affects light scattering too. Together with A.P. Levanyuk and A.A. Sobyanin, I spent considerable time and work on analyzing the problem; I believe we have ultimately clarified a great deal [54] but it would be impossible, and not really proper, to describe the resulting theory in this article.

Another optical problem that attracted much of my attention was how to take into account spatial dispersion in crystal optics. Some thirty years ago spatial dispersion, that is, the dependence of the dielectric permittivity (and, in the general case, of the tensor $\epsilon_{ij}(\omega, \boldsymbol{k})$) on the wave vector $\boldsymbol{k}$ of the wave, was totally ignored in courses on electromagnetic theory. Actually, L.D. Landau and E.M. Lifshits had clearly pointed out already, in the first edition of their *Electrodynamics of Continuous Media* (1957), that gyrotropy is indeed an effect of spatial dispersion. However, the second-order effects (relative to the ratio a/λ, where a is the atomic size and $\lambda = 2\pi/k$ is the

---

[7] From a general standpoint, my scientific biography ought to describe not only my work but also my 'school' and my 'teachers' who played their roles in my growing into a physicist (I am talking about myself but this remark is fairly general). Nevertheless, this is a very special topic, touched on in [151, 153]. I write there, among other aspects, about I.E. Tamm and L.D. Landau whom I consider to be my main 'teachers' (the words 'school' and 'teacher' are put in quotation marks because in this context their meaning is fuzzy and I do not like to use them).

wavelength), which are the only ones surviving in a nongyrotropic medium, are not mentioned (gyrotropy is an effect of the order of a/λ). In crystals these effects are indeed very small, although Lorentz had already mentioned them in the last century (the references are given in [55,56]). In 1958, being influenced by the discussion (or maybe I should rather say, in response to it) that appeared in the literature of the effects of second-order spatial dispersion (i.e., of the order of $(a/\lambda)^2$), based on model concepts, I analyzed the processes phenomenologically, by expanding the tensor $\epsilon_{ij}(\omega, \boldsymbol{k})$ or $\epsilon_{ij}^{-1}(\omega, \boldsymbol{k})$ in a series in powers of $\boldsymbol{k}$ up to terms of order k^2. This immediately reveals the optical anisotropy of cubic crystals (which is the effect that Lorentz had in mind). At the same time, it was pointed out in [57] that, even taking into account only first-order terms in a/λ in gyrotropic crystals, an 'additional' wave may arise close to an absorption line. As the whole matter of taking into account the spatial dispersion in crystal optics was 'ripe' at that moment, V.M. Agranovich and myself embarked on a systematic analysis of the problem, and to a certain extent completed it with the publication of a monograph [56] (the new edition is preparing now for publication). This book, as well as our paper [55], outlines the history of the evolution of crystal optics when spatial dispersion is taken into account and also in connection with the theory of excitons (see also [143]).

6.9 The Theory of Ferroelectric Phenomena. Soft Modes. The Limits of Applicability of the Landau Theory of Phase Transitions

In 1944–1945 a discovery was made at the P.N. Lebedev Physical Institute: the anomalous, very high, temperature-dependent dielectric permittivity ϵ of barium titanate $BaTiO_3$. The paucity of data and the polycrystallinity of the samples (a ceramic was studied) made it difficult to immediately realize that a new ferroelectric material had been found. As I worked (and still work) at the same institute, I got interested in the results of [58]. I knew Landau's theory of phase transitions [59] and thus easily constructed a phenomenological (thermodynamic) theory of ferroelectrics [60], and also came to the conclusion that $BaTiO_3$ was indeed a new ferroelectric. I need to remark that the Landau theory of phase transitions is a theory of a self-consistent (mean) field and in the simplest cases (say, for a single order parameter) coincides with the models used earlier (by Van der Waals, Weiss, and others). The main thing in the Landau theory is the generality of the approach and the consistency in satisfying symmetry constraints. However, it is also useful in simple situations because it works kind of automatically. In fact, I made use of this feature, although starting with phenomenological theories of ferroelectrics suggested earlier was also a possibility (see references in [60,61]).

In [60], the electric polarization P is used as a parameter, so that the thermodynamic potential near the second-order phase transition is written in the form

$$\Phi = \Phi_0 + \alpha P^2 + \frac{\beta}{2} P^4 - EP. \tag{6.8}$$

In the vicinity of the transition temperature $T = \Theta$, the coefficients are $\alpha = \alpha'_\Theta (T - \Theta)$, $\beta = \beta_\Theta$; if $T > \Theta$, the material is a paraelectric, and if $T < \Theta$, it is a pyroelectric, that is, it possesses a spontaneous polarization $P_0 \neq 0$, and

$$P_0^2 = -\alpha/\beta = \alpha'_\Theta (\Theta - T)/\beta_\Theta.$$

I shall remind the reader that a material with this property (or, more correctly, also a material in which a first-order phase transition occurs near the tricritical point) is what is known as a ferroelectric. Now, in a weak field we have

$$P = P_0 + \frac{\epsilon - 1}{4\pi} E,$$

and

$$\epsilon = 1 + \frac{2\pi}{\alpha'_\Theta (T - \Theta)} \quad \text{(for } T > \Theta\text{)},$$

$$\epsilon = 1 + \frac{\pi}{\alpha'_\Theta (\Theta - T)} \quad \text{(for } T < \Theta\text{)}. \tag{6.9}$$

Obviously, the '1' would be better replaced with ϵ_0, that is, with the permittivity not connected with the phase transition, or, better still, only the Curie–Weiss law $\epsilon \sim 1/|\Theta - T|$ should be used. The difference in a factor of 2 between the two formulas in (6.9) for $T > \Theta$ and $T < \Theta$ was sometimes called 'the law of 2' and was confirmed experimentally. A number of other formulas were obtained, and the data for some other, already known ferroelectrics were discussed in [60]. As for $BaTiO_3$, the structure of the pyroelectric (ferroelectric) phase was not yet known, and in [60] it was assumed to be tetragonal or rhombohedral. In both cases [60] gives diagrams of piezoelectric coefficients and emphasizes that at $T < \Theta$ not only pyroelectricity but also the piezoelectric effect appears in $BaTiO_3$.[8] In [60] I also considered the case of first-order phase transitions near the tricritical point (or, as we called it at the time, near the critical Curie point, in which the curve of a second-order phase transition on the p–T diagram changes to a curve of a first-order phase transition). This was achieved by adding a term $(\gamma/6)P^6$ to (6.8).

The polarization $\boldsymbol{P}$ is a vector, and if $\boldsymbol{P}$ plays the role of the order parameter, this parameter has in general three components. In Rochelle salt, which has a preferred axis even in its nonferroelectric phase, the order parameter

---

[8] A curio in this connection: in the 1950s I was a witness in a court of law in the USSR, upon the request of the American government, answering questions about the piezoeffect in $BaTiO_3$. The reason behind this was that someone in the USA claimed money in connection with the use of some $BaTiO_3$ piezoelements that the claimant had patented. The US government used my testimony (i.e., effectively my paper [60]) to dismiss the claim.

can be considered to have a single component: this is the polarization along the favored axis [62]. However, barium titanate in its paraelectric phase (i.e., above the temperature Θ) has cubic symmetry, so that one has to consider the vector $\boldsymbol{P}$. In this sense the theory in [60] was correct but limited: it could not determine the symmetry of ferroelectric phases. Alas, in 1945 I did not bother to extend the theory of phase transitions to a vector order parameter – because of the lack of experimental data, because of a heavy load of other work, and probably because I simply failed to think it through. I ultimately did it, though, after experimental results were obtained [62, 63]; I took into account elastic stress but only for second-order phase transitions, that is, I neglected terms of order P^6. The theory thus covered only transitions to the tetragonal or rhombohedral ferroelectric phase and did not give the solution for the orthorhombic phase. In this sense my work is less complete than the later publication of Devonshire [64] who, to be correct, took into account only one of the three possible terms of order P^6 (see [61, 65]). Unfortunately, and I have mentioned it before, papers by Soviet scientists had stopped being translated into English in the USSR[9] and our journals were not yet translated abroad; neither were we sending our papers for publication in the West. The consequences of this are obvious, but I do not wish to discuss questions of priority (this was partly done in [61] – that was an invited ('commissioned') talk, and written nearly forty years after the publications mentioned).

In addition to the aspects described above, the papers [62,63] did introduce the concept of a 'soft mode', which came into vogue some time later. Actually, the term 'soft mode' was not coined in [62, 63] and, furthermore, I failed to pay the subject the attention it deserved. The fact is, nevertheless, that this concept is ascribed in the literature to authors who did the work ten years later, and I believe that – at least in one case – the work was done in a way not a bit more complete than in [62,63]. The story is given in greater detail in [61]. We published an extensive discussion of the soft-mode aspects in [54] in connection with light-scattering problems.

Ferroelectrics are in many ways similar to ferromagnets, which is reflected in the similarity of these terms in English. I shall therefore mention two papers [66, 67] in which ferromagnets were considered in the vicinity of the Curie point, and in [67] I dealt with domain walls in which the magnetization changes in magnitude, not in direction. I shall also mention [68], which discussed the possibility of the existence of surface ferromagnetism.

To conclude this section, I shall outline the limits of applicability of the Landau theory of phase transitions. I have already emphasized (this is well known, though) that this was a mean-field theory, although it allowed one

---

[9] The publication of the excellent *Journal of Physics USSR* was stopped in 1947 as a consequence of the 'anti-cosmopolitanism' campaign. The termination was so abrupt that complete typeset issues were destroyed (for example, paper [32] contains a reference to its translation in the *Journal of Physics USSR*, even though the corresponding issue of this journal was never printed).

to calculate fluctuations of some quantities as long as they remained small. What lies behind this qualification? Obviously, if we calculate a quantity, say, the polarization in a ferroelectric, we can use the Landau theory as long as the condition

$$\overline{(\Delta P)^2} \ll P_0^2 \tag{6.10}$$

holds, that is, as long as the fluctuations of the polarization are small in comparison with the spontaneous polarization P_0 (I have mentioned that $P_0^2 = -\alpha/\beta$ if we use the potential (6.8); in (6.10), $\overline{(\Delta P)^2} = \overline{(P - P_0)^2}$, where the overbar indicates statistical averaging and, of course, $\overline{\Delta P} = 0$.) If this simple criterion is applied, it follows that the Landau theory is applicable if

$$\tau = \frac{\Theta - T}{\Theta} \gg \frac{k_B^2 \Theta \beta_\Theta^2}{32\pi^2 \alpha'_\Theta \delta^3} , \tag{6.11}$$

where δ is the coefficient of $(\nabla P)^2$ that must be added to the thermodynamic potential (6.8) when taking into account the nonuniformity of the order parameter (in this case, of the polarization P); furthermore, Θ in (6.11) is, as before, the transition temperature, $\alpha = \alpha'_\Theta (T - \Theta)$, and $k_B = 1.38 \times 10^{-16}$erg/K is the Boltzmann constant. Simple calculations that lead to inequality (6.11) are given in [69] and also in [61]; only that part of the fluctuation $\overline{(\Delta P)^2}$ which depends significantly on temperature T is singled out.

Note that the numerical coefficient $1/(32\pi^2)$ in the final expression was not spelled out in [69]. Other authors also act like this sometimes (see, e.g., [59, 70]), since a coefficient is not very important in the case of inequalities. The actual smallness of the coefficient does become important when we discuss specific transitions. The criterion (6.11), with the same or a different numerical coefficient, can be derived in ways that differ from the one outlined above (see Sects. 146 and 147 of [59], and [71]).

By 1960, publishers in the West (mostly in the USA) had started to translate Soviet journals; this may be the reason why [69] was often cited, and still is. The criterion (6.11) is even known as the 'Ginzburg criterion' and the quantity Gi $= k_B^2 \Theta \beta_\Theta^2 /(\alpha'_\Theta \delta^3)$ was called (perhaps in [70] for the first time) the Ginzburg number. It is of course flattering to have 'your own' criterion and number, but I never use this terminology. The point is not my modesty but rather the fact that in the Russian language (in contrast, I believe, to English) using your own surname in your own paper is an 'awkward' thing to do, it is not 'not done' (for the same reason, I never resort to the terms, quite widespread in the literature, 'Ginzburg–Landau theory' and 'Ginzburg–Pitaevskii theory').

Papers of interest in this context are a concrete discussion, based on the criterion (6.11), of the applicability of the Landau theory to various phase transitions (see [69,72] and certain papers, cited below, dealing with the theory of superfluidity of Helium II in the vicinity of the lambda point, and also the paper [73] on the theory of high-temperature superconductors).

6.10 The Superfluidity of Helium II Near the Lambda Point.
Other Publications on Superfluidity[10]

I remember Landau's theory of superfluidity [28] as one of the magnificent events in my life. This is indeed an exceptional paper. However, it was incomplete in several respects. I do not mean even the fact that Landau has later drastically changed [74] the excitation spectrum from the one he had originally chosen. It is more important that Landau did not consider the Bose statistics of ^{4}He atoms to be decisive for creating superfluidity. In fact, Feynman showed [75] that Bose statistics are important for superfluidity (actually, this was understood even before Feynman's work, after liquid ^{3}He was produced and manifested dramatic differences from liquid ^{4}He). Evidently, this could not affect the two-liquids hydrodynamics of Helium II constructed by Landau [28]. And finally – this is the central point for me here – Landau did not consider the region, near the lambda point, that is, the He II $\leftrightarrows$ He I transition. His quasi-microscopic approach is not valid in this region since the concentration of excitations (quasiparticles) grows too high and they do not form a gas any more. As for the hydrodynamic theory, it is based, among other things, on introducing the density of the superfluid component ρ_s of the liquid (He II), which is assumed to be some fixed function of p and T or of other thermodynamic variables. In reality a phase transition – in this case the lambda transition in helium – must be related to some order parameter η and its changes, and this parameter is not fixed from the start but is found from an equation, for example, the one following from the Landau theory of phase transitions. It is natural to assume that η (see below) is somehow related to ρ_s. As far as I know or can recall, L.D. Landau – the author of the theories of both superfluidity and phase transitions – was never interested in this problem; what is certain is that he never introduced an order parameter for He II. On the contrary, I got interested (as early as 1943) precisely in this aspect of the phase transition between He I and He II, this was covered in my first paper [76] which was devoted to superconductivity (I shall discuss this subject further on in the chapter). I should point out that no concrete result was reported in [76], and only a rather fuzzy hypothesis was formulated with respect to the possibility of a thermodynamic approach to calculating the critical velocity of the superfluid flow. This idea was to some extent elaborated on in [77], where the quantity ρ_s was chosen for the order parameter, and the thermodynamic potential $\Phi_{\mathrm{He\,II}} = \Phi_{\mathrm{He\,I}} + \alpha\rho_s + (1/2)\beta\rho_s^2 + (1/2)\rho_s v_s^2$ was used. It then follows that in equilibrium we have $\rho_s = \rho_{se} - v_s^2/2\beta$, where $\rho_{se} = |\alpha|/\beta$, that is, ρ_s depends on v_s and there exists a certain critical velocity at which $\rho_s = 0$. Actually, the note [77] mostly discusses a different

---

[10] The material of this and the next section of the paper is presented in more detail in Chap. 2 "Superconductivity and Superfluidity (What was Done and What was Not Done)" included in the present book.

explanation of how the critical velocity arises. All this is at best of historical significance and does not deserve a more detailed description. The same is true for the aspect of surface energy related to the tangential discontinuity of the velocity in Helium II [78]. Helium atoms stick to the walls, so that if the velocity of the superfluid flow is $v_\mathrm{s} \neq 0$, there must be a tangential-velocity discontinuity at the wall and it may seem that a rather considerable energy must be involved [78]. However, specially designed experiments proved that, with a high degree of accuracy this energy is zero [79]. This led to the hypothesis that $\rho_\mathrm{s} = 0$ at the wall surface, so that the flow $\rho_\mathrm{s} v_\mathrm{s}$ is also zero, and hence, that in the context of interest to us here, the velocity discontinuity is rather innocuous. It was the understanding of this feature that provided the stimulus for constructing the theory of superfluidity of Helium II near the lambda point, which I did together with L.P. Pitaevskii [80]. By that time the Ψ theory of superconductivity [81] had existed for some considerable time; the order parameter in this theory is the macroscopic wave function Ψ, and $|\Psi|^2 \sim n_\mathrm{s}$, where n_s is the concentration of 'superconducting' electrons. A function $\Psi = \eta e^{i\phi}$ was introduced, in a similar manner to that for helium, as the order parameter, and

$$\rho_\mathrm{s} = m\eta^2 = m|\Psi|^2 \,, \quad \boldsymbol{v}_\mathrm{s} = \frac{\hbar}{m}\nabla\phi \,, \tag{6.12}$$

where the mass of the helium atom ^{4}He must be chosen here as the mass m in the expression for $\boldsymbol{v}_\mathrm{s}$ (and can be chosen in the expression for ρ_s).

The thermodynamic potential of Helium II was written in the form

$$\Phi_{\mathrm{He\,II}} = \Phi_{\mathrm{He\,I}} + \frac{\hbar}{2m}|\nabla\Psi|^2 + \alpha|\Psi|^2 + \frac{\beta}{2}|\Psi|^4 \tag{6.13}$$

and we assumed that, as is usual in mean-field theory (Landau theory)

$$\alpha = \alpha'_{T_\lambda}(T - T_\lambda) \,, \quad \beta = \beta_{T_\lambda} = \mathrm{const} \,. \tag{6.14}$$

Furthermore, in accordance with the arguments given above, the following boundary condition at the wall (index 0) is assumed,

$$(\Psi)_0 = 0 \,; \tag{6.15}$$

in the Ψ theory of superconductivity, $\mathrm{d}\Psi/\mathrm{d}z = 0$ at the superconductor–vacuum boundary (z is the coordinate perpendicular to the boundary). One of the implications of (6.13) is that in Helium II at rest near the lambda point we have

$$\rho_\mathrm{se} = m|\Psi_\mathrm{e}|^2 = \frac{m\alpha'_{T_\lambda}(T_\lambda - T)}{\beta_{T_\lambda}} \,. \tag{6.16}$$

The theory makes it possible to solve a number of problems (the behavior of Helium II in capillaries and slits, the variation of ρ_s as $\boldsymbol{v}_\mathrm{s}$ increases, etc.). The recipe outlined in [80] was generalized to nonstationary processes by L.P. Pitaevskii [82].

The successful application of the Ψ theory of superconductivity [81] stimulated a belief that the Ψ theory of superfluidity [80] could be very efficient for analyzing the behavior of Helium II near the lambda point. However, this conclusion is wrong. The point is that the mean-field approximation is applicable nicely in the case of superconductors almost up to the critical point T_c. This is easily shown [69] by using inequality (6.11).

This is a good place to recall the physical meaning of the coefficient δ of the gradient term $(\nabla\eta)^2$ in the expression for the thermodynamic potential (this term, $\delta(\nabla\eta)^2$, is added to (6.8), where the parameter $\eta = P$ is chosen). It is immediately clear that with a nonuniform distribution of the order parameter, the characteristic distance – the coherence length over which the spatial distribution of η varies – is of the order $\xi \sim (\delta/\alpha)^{1/2}$; indeed, with this gradient of the order parameter we have $\delta(\nabla\eta)^2 \sim \delta\eta^2/\xi^2 \sim |\alpha|\eta^2$, that is, the 'correlation energy' $\delta(\nabla\eta)^2$ is of the order of the bulk energy $|\alpha|\eta^2$. The quantitative expression for ξ is obtained by considering the correlation function for the fluctuations of the parameter η; as the result we obtain $\xi^2 = 2\delta/\Phi_e''$, where $\Phi_e'' \equiv (\partial^2\Phi/\partial\eta^2)_e$ is the equilibrium value of the appropriate derivative. For a potential of type (6.8) and (6.13), we find $\eta_e = 0$ and $\Phi_e'' = 2\alpha = 2\alpha_{T_\lambda}'(T-T_\lambda)$ above the transition point, whence

$$\xi = \sqrt{\frac{\delta}{\alpha}} = \sqrt{\frac{\hbar^2}{2m\alpha_{T_\lambda}'(T_\lambda - T)}} = \frac{3.5 \times 10^{-8}}{\sqrt{T_\lambda - T}} \ (\text{cm}) \ , \qquad (6.17)$$

where known values of the coefficients for He II have been used (for details see [80,83–86]; the symbol η used above denotes both the order parameter and, for the Ψ theory, the modulus of the order parameter Ψ, but this is unlikely to lead to confusion). It is clear from (6.17) that the correlation length in liquid helium is large in comparison with the atomic size $a \sim 3 \times 10^{-8}$ cm (the interatomic distance in helium at $T = T_\lambda$ is $a = 3.57 \times 10^{-8}$ cm) only in the immediate vicinity of the lambda point. However, it is possible to show, using criterion (6.11), that the fluctuations there are already high and the entire approach of (6.13) and (6.14) cannot be used any more for certain quantitative calculations (see [83–86]). Running ahead of the story, I shall remark that the length ξ in ordinary superconductors is large (this is essentially caused by (6.14), including now not the helium atom mass m_{He} but the electron mass). Therefore the Ψ theory of superconductivity [81] usually has a wide scope of applicability.

The failure of the mean-field theory in liquid helium is manifested especially clearly in the fact that the density ρ_s near the lambda point follows not (6.16) but

$$\rho_{se} = 0.35\tau^\zeta \ \text{g cm}^{-3} \quad (\tau = (T_\lambda - T)/T_\lambda) \ , \qquad (6.18)$$

where $\zeta = 0.672 \pm 0.001$, or 2/3 for all practical purposes, in the range $10^{-6} \leq \tau \leq 10^{-2}$ ((6.16) implies $\zeta = 1$).

In view of this, the Ψ theory of superfluidity never gained much following. Furthermore, soon after it was published, the theory of phase transitions began its explosive development on the basis of the concepts of scale invariance of critical phenomena and on the basis of the field-theoretical approach using renormalization groups (see [59,70]). This was definitely a path to success but I still believe that the Ψ theory of superfluidity, after an appropriate generalization, remains useful and perhaps (I do not know the final answer) may prove sufficiently good for solving a number of problems. In fact, the Landau theory of phase transitions can be generalized (even if semiempirically) while retaining its general approach, by changing the temperature dependence (6.14) of the coefficients α, β, etc. in expressions of type (6.13). As far as I know, this approach was first suggested by Y.G. Mamaladze in 1967 [87]. A generalized theory was then discussed by a number of authors; A.A. Sobyanin and myself have been applying it for a number of years to analyze superfluidity close to the lambda point (see [83–86,144], where references can be found to our other papers as well). In this theory the potential Φ replacing (6.13) is written as

$$\Phi_{\text{He II}} = \Phi_{\text{He I}} + \frac{\hbar^2}{2m}|\nabla\Psi|^2 + \alpha|\Psi|^2 + \frac{\beta}{2}|\Psi|^4 + \frac{\gamma}{6}|\Psi|^6 , \qquad (6.19)$$

where

$$\alpha = -A\tau|\tau|^{1/3} , \quad \beta = \beta|\tau|^{2/3} , \quad \gamma = \text{const} , \quad \tau = (T_\lambda - T)/T_\lambda .$$

The coefficients in (6.19) are selected in such a way that (6.18) can hold for equilibrium Helium II with $\zeta = 2/3$. The equation for Ψ obtained from (6.19) allows one to solve a wide range of problems. The current status of the theory is described in [?, 84–86], of which [85, 144] are the easiest to access. For this reason, and for technical ones as well (insufficient space being one of them), I shall not go into any details (for the same reasons, the notation for the coefficients was slightly changed in (6.19) in comparison with [84–86] to simplify the expression). The fate of the generalized Ψ theory of superfluidity is not clear yet, since the amount of experimental data available for verifying it is painfully inadequate. If the theory is found to work with a wide spectrum of data to an accuracy of about 1% – and we hope it will – its use will be totally justified because more rigorous methods are incomparably more complex when applied to certain problems (dimensional effects and some others). In principle, almost any phenomenological theory finds itself in the same predicament. For instance, all problems in aerodynamics could be solved on the basis of the kinetic theory of gases. However, doing so in the range of applicability of the equations of hydrodynamics would be madness itself. The situations in crystal optics (see Sect. 6.8 and [56,143]) and in the Ψ theory of superconductivity are quite similar to this. I can explain for myself the insufficient attention to the current form of the Ψ theory of superfluidity [83–86,144] only by a suspicion that people wrongly identify it with the self-consistent field theory [80] (see above), and also as an effect of fashion and of misunderstanding the underlying physics.[11]

---

[11] I must point out that doubts were expressed [136] about the validity of the Ψ theory of superfluidity, at least in certain cases ([136] deals with surface tension

In addition to the Ψ theory of superfluidity near the lambda point, I discussed superfluidity in papers dealing with scattering of light (see Chap. 1 in this book; [52], critical velocities [77], the possible superfluidity of molecular hydrogen [88], superfluidity in neutron stars and in astrophysics in general [89]), and, finally, the thermomechanical circular-flow effect in a nonuniformly heated annular vessel containing a superfluid liquid (see Chap. 1 in this book; [90,91]). In this last case a circular flow of the superfluid component of liquid Helium II must appear in a nonuniformly heated annular (closed and non-simply connected) vessel filled with superfluid helium. Curiously, the idea of this effect grew [90] in connection with the thermoelectric effect in a superconducting circuit (see below). On the other hand, I came to the conclusion, which was earlier rejected, that there must exist thermoelectric phenomena in superconductors, many years earlier (than for superfluidity in [92]), using an analogy with the hydrodynamics of a superfluid liquid. The effect predicted in [90,91] has already been observed but, to the best of my knowledge, has attracted little attention even though it opens ample opportunities for studying superfluidity [91].

6.11 The Theory of Superconductivity

L.D. Landau's paper on the theory of superfluidity [28] ended with a discussion of the superconductivity problem. Superconductivity is considered to be analogous to superfluidity, which is perfectly correct, and is also connected with an energy gap in the spectrum of the 'electron liquid' in the metal. In a paper [76], mentioned earlier in this article, that I wrote in 1943 after the evacuation to Kazan, I postulated a certain spectrum of 'excitations' (i.e., quasiparticles, electrons, and holes) in a superconductor. In contrast to the spectrum in the normal state of the superconductor, a temperature-independent gap Δ was introduced for the superconducting state. Then the free energy, depth of penetration of the magnetic field, etc. were calculated for this spectrum. A comparison with the experimental data reported by 1940 led, for example for mercury, to the value $\Delta/(k_{\mathrm{B}}T_{\mathrm{c}}) = 3.1$. However, the Bardeen–Cooper–Schrieffer (BCS) microscopic theory of superconductivity constructed 14 years later (in 1957) gave, in agreement with the current experimental data, the value $2\Delta(0)/(k_{\mathrm{B}}T_{\mathrm{c}}) = 3.53$ in the case of weak coupling; furthermore, the gap $\Delta(T)$ is a function of temperature, and $\Delta(T_{\mathrm{c}}) = 0$. We see therefore that the quasi-microscopic model constructed in [76] was quite far from reality, even though it did contain some correct qualitative elements. This 'quasi-microscopic theory' of superconductivity was presented and somewhat extended in a monograph [93] and a review paper [94], but can now be of purely historical interest only.

---

of liquid helium in the vicinity of the lambda point). The criticism met with objections, so the problem as a whole appears to be open (see also [144]).

My second paper on the theory of superconductivity [92], written in the same year, 1943, had a different fate. It was assumed at the time (and was repeated many years later; see, e.g., [95]) that thermoelectric effects in the superconducting state are completely absent. This is not true, although in real systems the thermoelectric effects in a superconductor are in a certain sense small and difficult to observe. The thing is that both the superconducting current (with density j_s) and the normal current (with density j_n) carried by 'normal' electrons can flow through a metal in the superconducting state. Obviously, j_s and j_n are similar to the fluxes $\rho_s v_s$ and $\rho_n v_n$, respectively, in a superfluid liquid. The current j_n in a nonclosed superconductor (say, a rod) is not zero but in an isotropic material this current is compensated by the current j_s, so that the total current $j = j_s + j_n$ is zero. As a result, a thermoelectric current j_n results only in additional heat conduction. This factor was noted in [92] but the corresponding thermal-circulation heat conduction coefficient κ_c was not calculated: for that, one needed the microscopic theory of superconductivity, which had not yet been created. Such calculations, which were carried out many years later on the basis of the BCS theory, gave an estimate[12]

$$\kappa_c/\kappa_{el} \sim (k_B T_c/E_F)^2, \tag{6.20}$$

where κ_{el} is the thermal conduction coefficient due to 'normal' electrons in the superconductor and E_F is the Fermi energy in the metal under consideration. In ordinary – non-high-temperature – superconductors, $T_c \lesssim 10\,\mathrm{K}$ and $E_F \sim 10\,\mathrm{eV} \sim 10^5\mathrm{K}$, whence $\kappa_c/\kappa_{el} \lesssim 10^{-8}$. Actually, as mentioned above, this effect was only mentioned in [92]. It was pointed out, however, that no compensation of j_n and j_s happens in anisotropic superconductors (when the directions of ∇T and the crystal symmetry axes do not coincide), and also in spatially inhomogeneous superconductors. A side effect of [92] was a generalization of the Londons' superconductor electrodynamics, which was popular at that moment, to the anisotropic case. For a number of reasons, the observation of thermoelectric effects in the superconducting state is rather difficult and thus it stayed in the shadows. The first experimental data in this field was obtained only in 1974 (30 years after [92] was published!). In reality, the picture is still not quite clear, but here I shall only refer the reader to the review [96] and the papers [97–99, 137, 144–146] and make a remark with respect to high-temperature superconductors (HTSCs). In this case the estimate (6.20) yields $\kappa_c/\kappa_{el} \sim 10^{-2}$ (at $T \sim 100\,\mathrm{K}$ and $E_F \sim 0.1\,\mathrm{eV}$). It is possible for the thermal-circulation heat transfer in HTSCs to be even higher, since the estimate (6.20) is quite crude and does not hold for non-BCS-type superconductors. For this and a number of other reasons the study of thermoelectric effects in HTSCs must attract researchers [99, 145, 146]. In fact, this is also true for 'ordinary' superconductors. It is interesting that [145, 146] are my last research papers, published almost 55 years (!) after the paper [92] which started the discussion of the problem, and the problem is still 'alive' (see Article 1 in this book).

---

[12] This estimate was given only in [145], because some errors were made previously.

In those far-away years I continued working on some other aspects of superconductivity theory [93, 100–102], but here I only need to characterize [101] (for more details, see [144]). This paper showed that Londons' theory gives incorrect results for the magnetic field which destroys superconductivity in thin films and in calculating the surface energy σ_{ns} at the boundary between the superconducting and normal phases. To be precise, the Londons' theory can be 'saved' if we introduce a surface energy of the order of $\lambda H_c^2(8\pi)$, where λ is the penetration depth of the magnetic field into the superconductor and H_c is the thermodynamic critical magnetic field. The message was, therefore, the introduction (not the calculation) of a new parameter and, what is more important, one might expect that the surface energy would be of the order of $aH_c^2(8\pi)$, where $a \sim 10^{-8}$–10^{-7}cm is the atomic size, while in superconductors a is replaced with $\lambda \sim 10^{-5}$cm. The problem that faced us was thus to reveal the nature of the surface energy and calculate it. It was also understood, on the whole, that the Londons' theory would not work in strong fields H comparable with H_c. The problem as stated was solved in 1950 with the Ψ theory of superconductivity [81] that I mentioned above several times.

Breaking the historical logic of events, it will be easier to outline the main idea of the Ψ theory of superconductivity on the basis of the picture drawn in Sect. 6.10 for the Ψ theory of superfluidity. The difference lies in the fact that a current flowing through a superconductor is determined by the value of $|\Psi|^2$, proportional to the concentration of 'superconducting electrons'. In other words, we deal with the superfluidity of a charged liquid. In this connection, the thermodynamic potential (free energy) of a superconductor has the form (6.13) but with the gradient term replaced with

$$\frac{1}{2m}\left|-i\hbar\nabla\Psi - \frac{e}{c}A\Psi\right|^2, \tag{6.21}$$

where A is the vector potential of the magnetic field $H = \text{rot}\,A$; obviously, the energy of the field $H^2/(8\pi)$ must be added to (6.13). Since the Ψ theory of superconductivity withstood the test of time (in its range of applicability), it is widely used and presented in textbooks (e.g., see [103]). For this reason I need not go into greater detail and shall only add a few remarks. Some electrodynamic problems based on the Ψ theory were solved in [81]. In later papers I went into details of these problems and other aspects dealing with the behavior of thin films [104], inclusion of anisotropy [105], comparison with experimental data [106, 107], supercooling and superheating [108], ferromagnetic superconductors [109], quantization of magnetic flux [110], and some other problems [111]; I shall also mention the reviews [112, 144].

The story[13] connected with the determination of the charge e in (6.21) is rather curious (see [113, 114, 144]). I thought that this charge was not known

---

[13] In this place I (to some extent) repeat the information which we have placed in Chap. 2. It concerns the fact, that this article was written earlier than Chap. 2. But it seems to me that to modify the text of this article would be inexpedient. (*Author note to the present edition.*)

in advance and might have a certain effective value e_{ef}. With the data then available [106], a comparison with the experiment implied $e_{ef} = (2-3)e$, where e is the electron charge. Landau noted, however, that the introduction of an effective charge that may depend on coordinates is not admissible, since it violates the gradient invariance [106]. It is really striking that the idea never came to myself or anybody else that the charge e_{ef} in (6.21) might have some universal value, for example $2e$ (which is indeed what follows from the BCS theory and is confirmed by experiments [107]).

For many years superconductivity was a mysterious, unexplainable phenomenon and this was sufficient for attracting special attention to it. I realized then and see it even more clearly now that creating a microscopic theory of superconductivity was beyond my powers. However, following the events and thinking through the problem at the qualitative level was, of course, both possible and interesting. But the year 1957 saw the birth of the BCS theory and the veil of mystery slipped off. I accepted this with rather a mixture of frustration and relief. At any rate, I decided to give up working on superconductivity – I had numerous other plans. But this was not to be. First, certain 'tails' were left untied, or problems arose which I wished to solve and discuss [107, 108, 110, 111]. Then I could not resist getting interested in superconductivity in the cosmos [89, 115] and, finally, in 1964, I got excited about the problem of high-temperature superconductivity (HTSC), and continue to work on it now.

This is how it happened (see also [144, 147]). An idea came to mind of the possibility of surface superconductivity and, specifically, superconductivity of electrons in the surface (Tamm) levels. We discussed this aspect [116] in the spirit of the BCS theory. We did not think then about fluctuations. It was found out later that under certain conditions in the two-dimensional case (let alone the one-dimensional case) fluctuations destroy ordering. Now it is clear that two-dimensional superconductivity is in principle possible. It would, of course, be very desirable to have a dielectric inside and a superconductor on the surface. This problem is still on the wish list. The progress, however, went in a different direction. W. Little published a paper in which he discussed the possibility of a steep increase in the critical temperature T_c in a quasi-one-dimensional string, owing to the interaction of conduction electrons with bound electrons of the side 'branches' [117]. Unfortunately, the one-dimensional variant is fraught with shortcomings (large fluctuations, difficulties of implementation), and, as a result, but also independently of this argument, I immediately 'combined' [116] and [117] and suggested a two-dimensional variant of a high-temperature superconductor (a metal with a dielectric coating) [118]. This variant was later discussed in more detail [119] – the subject was the exciton mechanism of superconductivity[14] and the study

---

[14] The exciton mechanism is defined as mostly a BCS-type mechanism but with electron excitations – excitons – playing the role of phonons. In essence, [117, 118] implied precisely the exciton mechanism, although implicitly: the characteristic

of dielectric–metal–dielectric 'sandwiches' and layered structures. In the 1970s high-temperature superconductivity research was launched on a relatively broad scale at the Theoretical Physics Department of the P.N. Lebedev Physical Institute. The outcome of that effort was the world's first monograph on the subject [120] and a number of subsequent publications (see the volume of collected papers cited in [86]).

I believe that the activities aimed at high-temperature superconductivity [117–120] before it was actually discovered were both justifiable and useful. Layered (quasi-two-dimensional) materials were pointed out as potential candidates; it was shown that there was no ban in principle on values of $T_c \lesssim 300\,\mathrm{K}$, etc. At the same time, the theory was unable to provide any specific help in choosing a material, and no theory of HTSC (say, at the level of the BCS theory) was developed. Is there anything surprising in this? Of course not. It is enough to point out that several years have now passed since stable and reproducible high-temperature superconductors were synthesized[15] but the theory of such superconductors is still absent, and hot debates are raging (see [114, 121, 122, 138]). In this situation the publications [117–120] were often ignored; this still happens now. Well, this attitude can partly be understood: the saying goes that good intentions pave the road to hell. To be recognized and accepted, a theory must be sufficiently complete, and experimental results must be unequivocal and reliable; for example, a stable and reproducible HTSC material must be produced.

When HTSC materials were discovered, I rejoiced and tried to spread information about them [123]. Others will judge to what extent my HTSC work proved useful; as for me, I have always objected to priority demands and refuse to make such demands here (see [114] and Sect. 6.12).

It is only natural that we are still very interested in the HTSC problem and continue discussing it. I have already mentioned the results on the thermoelectric effect (see [145]). I regard constructing a macroscopic theory of high-T_c superconductors as an important achievement; the well-known HTSC materials with small coherence length [73] belong to the group for which this

---

energy of excitons E_{ex} is considerably higher than the phonon energy $\hbar\omega_{ph} \sim k\Theta_D$ (Θ_D is the Debye temperature). This is the factor that may increase the critical temperature T_c (for details, see [114, 119, 120]). Recently, other variants have been discussed in the framework of HTSC, where attraction between conduction electrons also stems not from phonons but from electrons in the system. Among other things, the role of spin excitations (spin waves) is discussed. To avoid confusion, it would be advisable to speak of the exciton mechanism only in connection with the 'electron' mechanism, in which the spin effects play an insignificant role.

[15] In my opinion, high-temperature superconductors should be defined as superconductors with $T_c > T_{b,N_2} = 77.4\,\mathrm{K}$ (T_{b,N_2} is the boiling temperature of liquid nitrogen at atmospheric pressure). Such materials (first, $YBa_2Cu_3O_{7-y}$ alloys) were obtained only at the beginning of 1987. However, materials with $T_c \gtrsim 30\,\mathrm{K}$, discovered in 1986, are also traditionally classified in the literature as high-temperature superconductors.

has been done. It proved possible in this case to combine the ordinary Ψ theory of superconductivity [81] for anisotropic materials [105] with the generalized Ψ theory which is valid in the critical region (see Sect. 6.10 and [83–87, 144]). Actually, this was done under the assumption that the order parameter is the scalar complex function Ψ. However, in HTSC, and in superconductors with 'heavy fermions', the order parameter may prove to be more complex (see [122]). I am not involved in microscopic theory; I only try to follow its evolution. The problems in which I am actively engaged nowadays are the macroscopic theory of HTSC, particularly for various order parameters, and the theory of thermoelectric effects. Of course, many other topics remain interesting and attractive.[16] Unfortunately, and I have already written about it in "Notes on the Occasion of My Jubilee" (p. 285 in [151]), working efficiently became difficult after I reached 65 (right now I am considerably older – I was born on October 4, 1916). To complicate things further, in April 1989 I was elected to the USSR parliament as a representative of the Academy of Sciences of the USSR and, until retiring from the parliament (see the article in the newspaper *Poisk*, No. 8, June 1989), had to devote considerable time to social issues.[17] This is why I am extremely skeptical about the prospects of further research for me. Nevertheless, I do not want to give up; I try to follow the progress in physics, and may – who knows – achieve something yet. Those who work in superconductivity have a nice objective, we could even say a dream, which is a very good stimulus for me. Before 1987 the dream was to create a high-temperature superconducting material ($T_c > T_{b,N_2} = 77.4\,\mathrm{K}$), and now it is to create a room-temperature superconductor (RTSC; $T_c \gtrsim 300\,\mathrm{K}$; see [138, 139, 144, 148]). The state of the art in this problem is nearly the same as it was for HTSC before 1986–1987 (see Chap. 3 in this book).

6.12 Concluding Remarks

In defiance of my initial intentions and expectations, this article turned out to be rather long. Consequently, as I was writing it, I attempted to cut out some bits, to cite almost exclusively reviews (and monographs), etc. As a result, a number of papers that I consider to deserve (or to have deserved) some attention were not characterized or even mentioned. These were papers on Rayleigh scattering of light in gases [125], sound dispersion in liquids [126] and dispersion relations in acoustics [127], investigation of stress by optical techniques [128], microwave radio emission (a proposal to use undulators) [129], the theory of electric fluctuations [130], the self-consistent theory of ferromagnets [131], the role of quantum fluctuations of the gravitational field [132], and the theory of van der Waals forces [133]. I could somewhat extend

---

[16] Among these, I see the problem of superdiamagnetism [124], which is for some reason ignored. On some touched questions see Chaps. 1–3 in this book.

[17] There was no need for me to retire, since the institute of the deputies of the USSR parliament was discontinued on January 1, 1992.

this list but this is hardly expedient, all the more so since everything that I had published before 1977 was cited rather fully in [26] (see also [149, 150]).

Is this attempt at writing a scientific biography successful? I am not a proper judge, but I doubt it. What I wrote seems to be more of an extended annotation to my papers. To be honest, the work is not devoid of interest; I looked up some old publications, and I summed it all up. The article will be of use to those who decide to write my obituary or my post-mortem biography.[18] What are the rest of the readers going to think? Who knows? But I find solace in the thought, already mentioned in the introduction, that the article can be skipped.

In conclusion, I wish to touch on priority and precedence questions. Some points were raised earlier in this article, and some in the article 'Who Created the Theory of Relativity?' (p. 217 in [151]). I have also mentioned priorities in the article about Landau (see p. 367 in [151]).

I am not a 'priority guard', although I usually notice whether authors refer to my work or not. As a rule, however, I am never offended if the references are absent. The point is, the physics literature has grown immensely and it is now impossible to quote every relevant paper. Neither is it possible to follow everything. People try to include references to reviews, to certain papers that are already 'in the cartridge clip', and so on. Only a few deliberately omit references that are due, but do such people deserve any heed?

Still, authors and speakers at conferences do face the 'priorities problem': whom to mention and whom to omit? In 1987 I gave an introductory talk at the International Conference on Cosmic Rays [50] in which I solved the 'priorities problem' in a radical manner: I mentioned almost no one and wrote no bibliographic references on the transparencies. To explain this, I argued that mentioning names only distracts the audience and, in addition, may irritate the unmentioned. In conclusion, I displayed a transparency with two sentences: 'Priority questions are a dirty business' and 'Priority mania, or supersensitivity, is an illness.'

This was my advice not to go too deeply into the priorities trap, but served in a jocular form. I describe it here since this ingredient was not included in the published text of the talk [50]. On another occasion, when I displayed this transparency to a different audience and in a different context, some participants misunderstood me [114]. However, what I have formulated above is indeed my belief, spoken in perfect honesty. Twice in my life I was involved in debates on priorities but I maintain that I was not defending mine:

---

[18] The Royal Society of London publishes fairly large post-mortem biographies of its British and foreign members (*Biographical Memories of Fellows of the Royal Society*). As Y.B. Zeldovich was such a foreign member, and so am I, I was asked to write the appropriate 'biographical memory'. This was a huge and hard task, in which I was saved and completely propped up by Zeldovich's volumes [1]; without them the 'memory' would have suffered and the amount of work would have been far greater. The article, written in 1988, was published in 1994 (see *Biographical Memories* **40**, 429, 1994).

I was motivated by my opponents' behavior, which I regarded as unacceptable (I explain this in more detail in [27, 43, 47, 55]). I am also dead against the procedure, which was unfortunately practiced in the USSR, of officially filing 'discoveries.' However, killing this purely bureaucratic perversion was not possible until 1992 (it is obvious that I have never submitted any claims for 'discoveries' and have not even applied for an invention certificate or patent, even though you cannot say anything against these; in this respect see [134]). With the disintegration of the USSR, the registration of discoveries seems to have petered out.

My advice, therefore, is not to let yourself slide into a priorities controversy. Timely publication of your results will normally guarantee the protection of your copyright in research. This is, however, completely true only in the conditions of openness and the speedy publication of papers, and generally of an efficient information exchange, which are now typical of the international scientific community. The losses, sometime enormous losses, that Soviet science suffered in the past were caused by obscurantism (we recall the fate of genetics and cybernetics in the USSR), by the termination of the publication of scientific journals in English (a good example was the *Journal of Physics USSR* – I have mentioned it earlier), and by various bureaucratic bans and restrictions which were imposed under the guise of guarding the secrets and priority of Soviet science (I mean the barriers against submitting papers and even sending offprints to the West, and so on and so forth). The only way to safeguard the normal progress of science and the rights and interests of scientists (and also their priority rights) is to throw out all these archaic, truly 'stagnation' phenomena. Add to this the need to carefully nurture the generally accepted norms of morality, that is, the 'moral conditioning' that the scientific environment requires as badly as our entire society.

While reading this paper once again to prepare the English-language edition, I remembered the book *The Problems of Theoretical Physics and Astrophysics* (a collection of papers for the 70th birthday of V.L. Ginzburg) published in 1989 (Nauka, Moscow). The collection consists of 48 papers by 73 authors. To my shame, I did not make use of the Russian 1992 and 1995 of the present paper editions to acknowledge once again the authors of the above-mentioned collection *The Problems of Theoretical Physics and Astrophysics* for their contributions. I am now taking the chance to do this – better late than never.

6.13 Notes

1*. This chapter was written in 1991 and was first published in the second edition of my book *On Physics and Astrophysics* (M.: Nauka, Fizmatlit, 1992). Then it appeared in the third edition of the book in 1995 and its English translation of 2001 (see [151]). In the present edition, the paper has undergone minor revision.

2*. It was only in 2001, 46 years after [60] was published, that I managed to attend an international conference devoted to ferroelectric phenomena (it was the 10th such conference; the talk [61] was submitted to the 6th such conference which was held in Japan in 1985). My talk, delivered at the 10th conference, was published [155]; it was close in content to [61], but supplemented that report. I note that report [61] with some comments to it placed in book [153], paper 5.

3*. As has already been mentioned in Note 1*, in the present edition the paper 'Scientific Autobiography – An Attempt' has not undergone serious revision. Meanwhile, summarizing here my activity I would like to make several comments.

As I repeatedly noted in various papers (in particular, in my 'Nobel' autobiography), my weak point is computational technique and the apparatus of theoretical physics. But it can't be helped, and I sometimes had to calculate literally with great effort. I remember particularly well the determination of the energy dependence of the cross section for photon scattering by particle with spins $(1/2$–$3/2)$. This happened in the hard times of 1941 and 1942, and I was eager to somehow complete the clarification of the issue of whether or not the introduction of a higher spin state (spin $3/2$) leads to the cross-section 'cut-off.' I used the 'correspondence method', and it looked somehow unreliable. It fortunately turned out that I had not been mistaken, and it was further confirmed by V.Y. Fainberg (see [30, 33, 34]). As I have already noted in Sect. 6.5, I generally retreated from the study of potentialities in the field of relativistic theory of particles with internal degrees of freedom, that is, as a matter of fact, non-point particles because of the mathematical difficulties (see also [32–34]). Unwillingness to carry out calculations strengthened as I grew older, and I generally longed for other things such as history, publicism, and so on, which had already been predicted by L.I. Mandelshtam as far back as 1943 or 1944 (see [159] and p. 360 in [151]).

When I had students and post-graduates, some problems could be offered to them. They sometimes solved these problems, and in some cases I also participated. Then those were our collective works. But I shall permit myself to state that I have never dictated my will on anyone and never made use of 'youth labor power.' I have never been able to do that, although, to tell the truth, sometimes I would have liked it. As a result, I had not a single joint work, for example, with such talented, former post-graduate students as E.S. Fradkin and L.V. Keldysh. At the same time, in different papers I suggested many times various subjects and problems for possible study, but usually quite in vain. I will emphasize that I do not mean some generally known important problems of a principle character because everyone knows that there are such problems in physics and, say, in cosmology. But their allusion will of course give nothing, except that it may serve as another stimulus for a young man. Here I mean concrete problems, sometimes minor and particular. As an example, I will mention the quantum consideration of a transverse electromagnetic field carried away by the charge, touched upon both in my autobiography and in Sect. 6.2. Some problems concerning superconductivity and superfluidity were mentioned in my Nobel Lecture, while others were discussed in my papers and books [2,22,36,56]. Moreover, most of the problems were again listed in preprint [160]. Some of this may perhaps be used by students and post-graduates, in particular, as subjects of research. (See 5*.)

It remains to note that I wrote a whole number of papers devoted not only purely to various scientific issues, but also politics, the history of science, the struggle with pseudoscience, attitudes to religion, etc. These papers, the same as references to other works, can be found in books [151, 153]. Furthermore, some papers are laid out in the site of *Usp. Fiz. Nauk*: www.ufn.ru, home page "Physics Uspekhi Tribune." (See also my small book [170].)

4*. While preparing the English version of this book I thought that I was wrong not to have paid attention to this point (I mean papers [161, 162]). First, it was shown in these works that when a star collapses and its radius becomes close to the Schwarzschild radius, the magnetic moment of the star tends to zero. This was the first indication of the effect which further on was described by the phrase 'black holes have no hair.' Moreover, in [161], several years before the discovery of neutron stars (I think that the revelation of pulsars in 1968 can be thought of as such a discovery), it had been stated that the magnetic field of neutron stars can reach huge values of 10^{12}–10^{13} Oe. I do not know if this circumstance, which was further confirmed, had been pointed out before me, which is quite possible. But I am writing about it as an illustration of the following: the most important thing is sometimes to ask a question and the answer, even in the case of obtaining an important result, is more or less immediately obvious. Indeed, a star, even a neutron star, is very well-conducting. That is why we may think that in a collapse of an 'ordinary' star the magnetic flux is conserved and, thus, in a rough approximation $H_0 r_0^2 \sim H_n r_n^2$, where H_0 and H_n are the characteristic fields for an ordinary and neutron stars and r_0 and r_n are the radii of these stars. Hence, if, for example, $r_0 \sim 10^6$ km (the Sun radius is $r_0 = 7 \cdot 10^5$ km) and $r_n \sim 10$ km and $H_0 \sim 100$ Oe, then the field is $H_n \sim 10^{12}$ Oe.

5*. Preprint [160] was included in the book *Seminar* [163] prepared at the Theoretical Department of the P.N. Lebedev Physical Institute (mainly due to the efforts of Y.M. Bruk and B.M. Bolotovskii) on the occasion of my 90th birthday and published in Fizmatlit (I am grateful for that to Y.M. Bruk, B.M. Bolotovskii, and especially Larisa Alekseevna Panyushkina). As was said at the end of my 'Nobel' autobiography (Chap. 5 in this book), in connection with this jubilee three scientific sessions were held, and their content is reflected in a number of papers [164–169] which also present some results obtained by me and mentioned in the present paper.

Finally, three special sessions were held in connection with my 90th birthday (two of them in Moscow, and one in Nizhny Novgorod); the reports given in these scientific sessions were published in three issues of the journal *Uspekhi Fizicheskikh Nauk* [*Phys.-Usp.* **50**, No. 3, p. 293; No. 4, p. 331; and No. 5, p. 529

(2007)]. In addition, three special issues (Festschrift) of the scientific journals (celebrating my anniversary date) *Uspekhi Fizicheskikh Nauk* [*Physics-Uspekhi* **49**, No. 10 (2006)], *Journal of Superconductivity and Novel Magnetism* **19**, Nos. 3–5, July 2006 (Dr. Vladimir Kresin editor, and Dr. Ivan Božović guest editor) and *Ferroelectrics* **354** (2007): a special issue dedicated to Professor Vitali L Ginzburg in celebration of his 90th birthday (guest editors Vladimir Fridkin, Stephen Ducharme, Wolfgang Kleemann, and Yoshiro Ishibashi) were brought out. I wish to express my profound gratitude to all the authors of these articles and the editors, and to everybody who participated in these actions.

References

1. Ya.B. Zeldovich, *Izbrannye Trudy: Khimicheskaya Fizika i Gidrodinamika*, Nauka, Moscow, 1984 (in Russian) [*Selected Works: Chemical Physics and Hydrodynamics*, ed. by J.P. Ostriker, Princeton University Press, Princeton, 1992]; *Chastitsy, Yadra, Vselennaya*, Nauka, Moscow, 1985 (in Russian) [*Particles, Nuclei, and the Universe*, ed. by J.P. Ostriker, Princeton University Press, Princeton, 1993].
2. V.L. Ginzburg, *Teoreticheskaya Fizika i Asrofizika*, 3rd edn., Moscow, Nauka, 1987 (in Russian). [*Applications of Electrodynamics in Theoretical Physics and Astrophysics*, 3rd edn., Gordon and Breach, New York, 1989].
3. V.A. Fock, Sow. Phys. **6**, 425, 1934.
4. A.A. Smirnov, Zh. Eksp. Teor. Fiz. **5**, 687, 1935.
5. V.L. Ginzburg, Dokl. Akad. Nauk SSSR **23**, 773, 1939.
6. W. Heitler, *The Quantum Theory of Radiation*, Inostrannaya Literatura, Moscow, 1956 (in Russian). [*The Quantum Theory of Radiation* (International Series of Monographs on Physics), 1st edn., Clarendon Press, Oxford, 1936].
7. V.L. Ginzburg, Dokl. Akad. Nauk SSSR **23**, 896, 1939.
8. V.L. Ginzburg, Dokl. Akad. Nauk SSSR **24**, 130, 1939.
9. V.L. Ginzburg, Usp. Fiz. Nauk **140**, 687, 1983 [Sov. Phys.–Uspekhi **26**, 713, 1983].
10. V.L. Ginzburg, Zh. Eksp. Teor. Fiz. **9**, 981, 1939.
11. I.E. Tamm and I.M. Frank, Dokl. Akad. Nauk SSSR **14**, 107, 1937.
12. L.D. Landau and E.M. Lifshitz, *Elektrodinamika Sploshnykh Sred*, Vol. 8, Fizmatlit, Moscow, 1982. [*Electrodynamics of Continuous Media*, Pergamon, Oxford, 1984.]
13. V.L. Ginzburg, Zh. Eksp. Teor. Fiz. **10**, 601, 1940.
14. L.D. Landau and E.M. Lifshitz, *Teoriya Polya*, Vol. 2, Nauka, Moscow, 1988. [*The Classical Theory of Fields*, Pergamon, Oxford, 1975.]
15. V.L. Ginzburg, Zh. Eksp. Teor. Fiz. **10**, 608, 1940.
16. B.M. Bolotovskii, Usp. Fiz. Nauk **62**, 201, 1957.

---

[18] References are to publications in Russian, but sometimes information on the existing translations is included. As for those papers whose translations are not indicated, the reader should remember that *ZhETF* (*JETP*), *DAN SSSR* (*Doklady AN SSSR*) and some others were, and still are, translated into English, except for a break from roughly 1947 to the mid-1950s.

17. V.L. Ginzburg, Tr. Fiz. Inst. Akad. Nauk SSSR **176**, 3, 1986 (in Russian). A slightly abridged version of this article was published in *The Lesson of Quantum Theory*, ed. by J. de Boer et al., Elsevier, New York, 1986, p. 113; V.L. Ginzburg, in *Progress in Optics*, ed. by E. Wolf, Vol. 32, Elsevier, Amsterdam, 1993, p. 267.

18. V.L. Ginzburg, Zh. Eksp. Teor. Fiz. **10**, 589, 1940.

19. V.L. Ginzburg and I.M. Frank, Dokl. Akad. Nauk SSSR **56**, 583, 1947.

20. V.L. Ginzburg and V.P. Frolov, Usp. Fiz. Nauk **153**, 633, 1987 [Sov. Phys.–Uspekhi **30**, 1073, 1987]; Tr. Fiz. Inst. Akad. Nauk SSSR **197**, 8, 1989; Phys. Lett. A **116**, 423, 1986.

21. V.L. Ginzburg and I.M. Frank, Zh. Eksp. Teor. Fiz. **16**, 15, 1946; J. Phys. USSR **9**, 353, 1945.

22. V.L. Ginzburg and V.N. Tsytovich, *Perekhodnoye Izlucheniye i Perekhodnoye Rasseyaniye*, Nauka, Moscow, 1984 (in Russian) [*Transition Radiation and Transition Scattering*, Hilger, Bristol, 1990].

23. V.L. Ginzburg and V.N. Tsytovich, Zh. Eksp. Teor. Fiz. **88**, 84, 1985 [Sov. Phys.–JETP **61**, 48, 1985]; see also V.L. Ginzburg, Radiophysics **28**, 1211, 1985.

24. V.L. Ginzburg, Zh. Eksp. Teor. Fiz. **12**, 449, 1942; J. Phys. USSR **6**, 167, 1942.

25. I.E. Tamm and V.L. Ginzburg, Izv. Akad. Nauk SSSR, Ser. Fiz. **7**, 30, 1943.

26. *Vitaly Lazarevich Ginzburg* (Bibliography of USSR Scientists, 'Physicists' series, issue 21), Nauka, Moscow, 1978 (in Russian).

27. V.L. Ginzburg, Ann. Rev. Astron. Astrophys. **28**, 1, 1990.

28. L.D. Landau, Zh. Eksp. Teor. Fiz. **11**, 592, 1941; J. Phys. USSR **5**, 71, 1941.

29. V.L. Ginzburg, Zh. Eksp. Teor. Fiz. **11**, 620, 1941; J. Phys. USSR **5**, 47, 1941; Dokl. Akad. Nauk SSSR **31**, 319, 1941.

30. V.L. Ginzburg, Zh. Eksp. Teor. Fiz. **13**, 33, 1943; J. Phys. USSR **8**, 33, 1944; Phys. Rev. **63**, 1, 1943.

31. H.J. Bhabha, Phil. Mag. **43**, 33, 1952.

32. V.L. Ginzburg and I.E. Tamm, Zh. Eksp. Teor. Fiz. **17**, 227, 1947.

33. V.L. Ginzburg and V.I. Man'ko, Fiz. Elem. Chastits At. Yadra **7**, 3, 1976 [Sov. J. Part. Nucl. **7**, 1, 1976].

34. V.L. Ginzburg, *Quantum Field Theory and Quantum Statistics (in Honour of E.S. Fradkin)*, Hilger, Bristol, Vol. 2, p. 15, 1987.

35. V.L. Ginzburg, *Teoriya Rasprostraneniya Radiovoln v Ionosfere* [*Theory of Propagation of Radio Waves in the Ionosphere*], Gostekhizdat, Moscow, 1949 (in Russian).

36. V.L. Ginzburg, *Rasprosraneniye Elektromagnitnykh Voln v Plazme*, 2nd edn., Nauka, Moscow, 1967 (in Russian). The first edition was published in 1960. There are three English translations; the last and best one is *The Propagation of Electromagnetic Waves in Plasmas*, Pergamon, Oxford, 1970.

37. V.L. Ginzburg, Zh. Eksp. Teor. Fiz. **13**, 149, 1943; J. Phys. USSR **7**, 289, 1943.

38. V.L. Ginzburg, Zh. Eksp. Teor. Fiz. **21**, 788, 1951.

39. V.L. Ginzburg, Dokl. Akad. Nauk SSSR **35**, 302, 1942.

40. V.L. Ginzburg and A.A. Rukhadze, *Volny v Magnitoaktivnoi Plazme* (*Waves in Magnetoactive Plasma*), Nauka, Moscow, 1975 (in Russian). The first edition was published in 1970; English translation in *Handbuch der Physik*, Vol. 49/4, p. 395, Springer, Berlin, 1972.

41. V.L. Ginzburg, Tr. Fiz. Inst. Akad. Nauk SSSR **18**, 55, 1962.
42. V.L. Ginzburg, Dokl. Akad. Nauk SSSR **52**, 491, 1946.
43. V.L. Ginzburg, *The Early Years of Radioastronomy*, ed. by W.T. Sullivan, p. 289, Cambridge University Press, Cambridge, 1984.
44. V.L. Ginzburg, Priroda **10**, 80, 1986.
45. H. Alfven and N. Herlofson, Phys. Rev. **78**, 616, 1950; K.O. Kipenheuer, Phys. Rev. **79**, 738, 1950.
46. V.L. Ginzburg, Dokl. Akad. Nauk SSSR **76**, 377, 1951.
47. V.L. Ginzburg, *Early Years of Cosmic Ray Studies*, ed. by Y. Secido and H. Elliot, p. 411, Reidel, Dordrecht, 1985.
48. V.L. Ginzburg and S.I. Syrovatskii, *Proiskhozhdeniye Kosmicheskikh Luchei*, Izd. Akad. Nauk SSSR, Moscow, 1963 (in Russian). A more widely known version is the enlarged English translation *Origin of Cosmic Rays*, Pergamon, Oxford, 1964.
49. *Astrofizika Kosmicheskikh Luchei*, ed. by V.L. Ginzburg, Nauka, Moscow, 1984 (in Russian) (2nd edn.: Nauka, Moscow, 1990) [V.S. Berezinskii, S.V. Bulanov, V.A. Dogiel, V.L. Ginzburg, and V.S. Ptuskin, *Astrophysics of Cosmic Rays*, Elsevier Science, Amsterdam, 1990].
50. V.L. Ginzburg, Usp. Fiz. Nauk **155**, 185, 1988 [Sov. Phys.–Uspekhi **31**, 491, 1988].
51. V.L. Ginzburg and V.A. Dogel, Usp. Fiz. Nauk **158**, 3, 1989 [Sov. Phys.–Uspekhi **32**, 385, 1989]; Space Sci. Rev. **49**, 311, 1989.
52. V.L. Ginzburg, Zh. Eksp. Teor. Fiz. **13**, 243, 1943; J. Phys. USSR **7**, 305, 1943.
53. V.L. Ginzburg, Izv. Akad. Nauk SSSR, Ser. Fiz. **9**, 174, 1945; Zh. Eksp. Teor. Fiz. **34**, 246, 1958 [Sov. Phys.–JETP **34**, 170, 1958].
54. V.L. Ginzburg, A.P. Levanyuk, and A.A. Sobyanin, Usp. Fiz. Nauk **130**, 615, 1980 [Phys. Rep. **57**, 152, 1980]; see also article in *Light Scattering Near Phase Transitions* (Modern Problems in Condensed Matter Physics series), Vol. 5, p. 3, North-Holland, Amsterdam, 1983.
55. V.L. Ginzburg, Phys. Rev. **194**, 245, 1990.
56. V.M. Agranovich and V.L. Ginzburg, *Kristallooptika s Uchetom Prostranstvennoi Dispersii* [*Crystal Optics with Spatial Dispersion*], 2nd edn., Nauka, Moscow, 1979 (in Russian). The first edition was printed in 1965, the English translation in 1966 [translation of 2nd edition: *Crystal Optics with Spatial Dispersion and Excitons*, Springer, Berlin, Heidelberg, 1984].
57. V.L. Ginzburg, Zh. Eksp. Teor. Fiz. **34**, 1593, 1958 [Sov. Phys.–JETP **34**, 1096, 1958].
58. B.M. Vul and I.M. Goldman, Dokl. Akad. Nauk SSSR **49**, 154, 177, 1945.
59. L.D. Landau and E.M. Lifshitz, *Statisticheskaya Fizika*, Vol. 5, Part 1, Nauka, Moscow, 1976. [*Statistical Physics*, Pergamon, Oxford, 1980.]
60. V.L. Ginzburg, Zh. Eksp. Teor. Fiz. **15**, 739, 1945; J. Phys. USSR **10**, 107, 1946.
61. V.L. Ginzburg, Tr. Fiz. Inst. Akad. Nauk SSSR **180**, 3, 1987; Ferroelectrics **76**, 3, 1987.
62. V.L. Ginzburg, Usp. Fiz. Nauk **38**, 490, 1949.
63. V.L. Ginzburg, Zh. Eksp. Teor. Fiz. **19**, 36, 1949.
64. A. Devonshire, Phil. Mag. **40**, 1040, 1949; **42**, 1065, 1951.
65. M.Ya. Shirobokov and L.P. Kholodenko, Zh. Eksp. Teor. Fiz. **21**, 1237, 1250, 1951.

66. V.L. Ginzburg, Zh. Eksp. Teor. Fiz. **17**, 833, 1947.
67. L.N. Bulaevskii and V.L. Ginzburg, Zh. Eksp. Teor. Fiz. **45**, 772, 1963 [Sov. Phys.–JETP **18**, 530, 1964]; Pis'ma Zh. Eksp. Teor. Fiz. **11**, 404, 1970 [JETP Letters **11**, 272, 1970].
68. L.N. Bulaevskii and V.L. Ginzburg, Fiz. Met. Metalloved. **17**, 631, 1964.
69. V.L. Ginzburg, Fiz. Tverd. Tela **2**, 2031, 1960 [Sov. Phys. Solid State **2**, 1824, 1960].
70. A.Z. Patashinskii and V.L. Pokrovskii, *Fluktuatsionnaya Teoriya Fazovykh Perekhodov*, Nauka, Moscow, 1982 (in Russian) [English translation of 1st edition: *Fluctuation Theory of Phase Transitions*, ed. by P.J. Shepherd (Pergamon, Oxford 1979)].
71. A.P. Levanyuk, Zh. Eksp. Teor. Fiz. **36**, 810, 1959 [Sov. Phys.–JETP **36**, 571, 1959].
72. V.L. Ginzburg et al., Ferroelectrics **73**, 171, 1987.
73. L.N. Bulaevskii, V.L. Ginzburg, and A.A. Sobyanin, Zh. Eksp. Teor. Fiz. **94**, 355, 1988 [Sov. Phys.–JETP **68**, 1499, 1988]; Physica C **152**, 378, 1988; Physica C **153–155**, 1617, 1988.
74. L.D. Landau, J. Phys. USSR **11**, 91, 1947.
75. R.P. Feynman, *Statisticheskaya Mekhanika*, Mir, Moscow, 1978 (in Russian) [*Statistical Mechanics. A Set of Lectures*, W.A. Benjamin, Reading, Massachusetts, 1972; the original paper in question was written earlier: R.P. Feynman, Phys. Rev. **91**, 1291, 1301, 1953; Phys. Rev. **94**, 262, 1954].
76. V.L. Ginzburg, Zh. Eksp. Teor. Fiz. **14**, 134, 1944.
77. V.L. Ginzburg, Dokl. Akad. Nauk SSSR **69**, 161, 1949.
78. V.L. Ginzburg, Zh. Eksp. Teor. Fiz. **29**, 244, 1955 [Sov. Phys.–JETP **2**, 170, 1955].
79. G.A. Gamtsemlidze, Zh. Eksp. Teor. Fiz. **34**, 1434, 1958 [Sov. Phys.–JETP **34**, 992, 1958].
80. V.L. Ginzburg and L.P. Pitaevskii, Zh. Eksp. Teor. Fiz. **34**, 1240, 1958 [Sov. Phys.–JETP **34**, 858, 1958].
81. V.L. Ginzburg and L.D. Landau, Zh. Eksp. Teor. Fiz. **20**, 1054, 1950.
82. L.P. Pitaevskii, Zh. Eksp. Teor. Fiz. **35**, 408, 1958 [Sov. Phys.–JETP **35**, 282, 1959].
83. V.L. Ginzburg and A.A. Sobyanin, Usp. Fiz. Nauk **120**, 153, 1976 [Sov. Phys.–Uspekhi **19**, 773, 1976].
84. V.L. Ginzburg and A.A. Sobyanin, Low Temp. Phys. **49**, 507, 1982.
85. V.L. Ginzburg and A.A. Sobyanin, Usp. Fiz. Nauk **154**, 545, 1988 [Sov. Phys.–Uspekhi **31**, 289, 1988]; Japan J. Appl. Phys. **26**, Suppl. 26-3, Part 3, 1785, 1987.
86. V.L. Ginzburg and A.A. Sobyanin, in *Superconductivity, Superdiamagnetism, Superfluidity*, ed. by V.L. Ginzburg, p. 242, Mir, Moscow, 1987.
87. Yu.G. Mamaladze, Zh. Eksp. Teor. Fiz. **52**, 729, 1967 [Sov. Phys.–JETP **25**, 479, 1967].
88. V.L. Ginzburg and A.A. Sobyanin, Pis'ma Zh. Eksp. Teor. Fiz. **15**, 343, 1972 [JETP Lett. **15**, 242, 1972].
89. V.L. Ginzburg, Usp. Fiz. Nauk **97**, 601, 1969 [Sov. Phys.–Uspekhi **12**, 241, 1969]; J. Stat. Phys. **1**, 3, 1969.
90. V.L. Ginzburg, G.F. Zharkov, and A.A. Sobyanin, Pis'ma Zh. Eksp. Teor. Fiz. **20**, 223, 1974 [JETP Lett. **20**, 97, 1974].

91. V.L. Ginzburg and A.A. Sobyanin, Zh. Eksp. Teor. Fiz. **85**, 1606, 1983 [Sov. Phys.–JETP **56**, 934, 1984].

92. V.L. Ginzburg, Zh. Eksp. Teor. Fiz. **14**, 177, 1944; J. Phys. USSR **8**, 148, 1944.

93. V.L. Ginzburg, *Sverkhprovodimost'* [*Superconductivity*], Izd. Akad. Nauk SSSR, Moscow, 1946 (in Russian).

94. V.L. Ginzburg, Usp. Fiz. Nauk **48**, 26, 1952; Fortsch. Phys. **1**, 101, 1953.

95. A.C. Rose-Innes and E.H. Rhoderick, *Vvedenie v Fiziku Sverkhprovodimosti*, Mir, Moscow, 1972 [*Introduction to Superconductivity* (International Series of Monographs on Solid State Physics, Vol. 6), Pergamon, Oxford, 1969].

96. V.L. Ginzburg and G.F. Zharkov, Usp. Fiz. Nauk **125**, 19, 1978 [Sov. Phys.–Uspekhi **21**, 381, 1978].

97. V.L. Ginzburg, G.F. Zharkov, and A.A. Sobyanin, J. Low Temp. Phys. **47**, 427, 1982; J. Low. Temp. Phys. **56**, 195, 1984.

98. G.F. Zharkov, in *Superconductivity, Superdiamagnetism, Superfluidity*, ed. by V.L. Ginzburg, p. 126, Mir, Moscow, 1987.

99. V.L. Ginzburg, Pis'ma Zh. Eksp. Teor. Fiz. **49**, 50, 1989 [JETP Lett. **49**, 58, 1989]; for more detail, see J. Superconductivity **2**, 323, 1989; Usp. Fiz. Nauk **161**, 1, 1991 [Sov. Phys.–Uspekhi **34**, 101, 1991].

100. V.L. Ginzburg, Zh. Eksp. Teor. Fiz. **14**, 326, 1944.

101. V.L. Ginzburg, Zh. Eksp. Teor. Fiz. **16**, 87, 1946; J. Phys. USSR **9**, 305, 1945.

102. V.L. Ginzburg, J. Phys. USSR **11**, 93, 1947.

103. E.M. Lifshitz and L.P. Pitaevskii, *Statisticheskaya Fizika*, Vol. 9, Part 1, Nauka, Moscow, 1978. [*Statistical Physics*, Pergamon, Oxford, 1980.]

104. V.L. Ginzburg, Dokl. Akad. Nauk SSSR **83**, 385, 1952; Dokl. Akad. Nauk SSSR **118**, 464, 1958.

105. V.L. Ginzburg, Zh. Eksp. Teor. Fiz. **23**, 236, 1952.

106. V.L. Ginzburg, Zh. Eksp. Teor. Fiz. **29**, 748, 1955.

107. V.L. Ginzburg, Zh. Eksp. Teor. Fiz. **36**, 1930, 1959 [Sov. Phys.–JETP **36**, 1372, 1959].

108. V.L. Ginzburg, Zh. Eksp. Teor. Fiz. **34**, 113, 1958 [Sov. Phys.–JETP **34**, 78, 1958].

109. V.L. Ginzburg, Zh. Eksp. Teor. Fiz. **31**, 202, 1956 [Sov. Phys.–JETP **4**, 153, 1957].

110. V.L. Ginzburg, Zh. Eksp. Teor. Fiz. **42**, 299, 1962 [Sov. Phys.–JETP **15**, 207, 1962].

111. V.L. Ginzburg, Dokl. Akad. Nauk SSSR **110**, 358, 1956; Zh. Eksp. Teor. Fiz. **30**, 593, 1956; Zh. Eksp. Teor. Phys. **31**, 541, 1956; Zh. Eksp. Teor. Fiz. **44**, 2104, 1963 [Sov. Phys.–JETP **3**, 621, 1956; Sov. Phys.–JETP **4**, 594, 1957; Sov. Phys.–JETP **17**, 1415, 1963]; Physica **24**, 42, 1958.

112. V.L. Ginzburg, Nuovo Cim. **2**, 1234, 1955.

113. V.L. Ginzburg, Usp. Fiz. Nauk **94**, 181, 1968 [Sov. Phys.–Uspekhi **11**, 135, 1968]; see also Phys. Today **42** (5), 54, 1989.

114. V.L. Ginzburg, Progr. Low Temp. Phys. **12**, 1, 1989.

115. V.L. Ginzburg and D.A. Kirzhnits, Zh. Eksp. Teor. Fiz. **47**, 2006, 1964 [Sov. Phys.–JETP **20**, 1346, 1965].

116. V.L. Ginzburg and D.A. Kirzhnits, Zh. Eksp. Teor. Fiz. **46**, 397, 1964 [Sov. Phys.–JETP **19**, 269, 1964]; see also V.L. Ginzburg, Phys. Scripta **T27**, 76, 1989.

117. W.A. Little, Phys. Rev. A **134**, 1416, 1964.
118. V.L. Ginzburg, Zh. Eksp. Teor. Fiz. **47**, 2318, 1964 [Sov. Phys.–JETP **20**, 1549, 1965]; Phys. Lett. **13**, 101, 1964.
119. V.L. Ginzburg, Usp. Fiz. Nauk **95**, 91, 1968; Usp. Fiz. Nauk **101**, 185, 1970; Usp. Fiz. Nauk **118**, 35, 1976 [Contemp. Phys. **9**, 355, 1968; Sov. Phys.–Uspekhi **13**, 335, 1971; Sov. Phys.–Uspekhi **19**, 174, 1976]; Pis'ma Zh. Eksp. Teor. Fiz. **14**, 572, 1971 [JETP Lett. **14**, 396, 1971]; Ann. Rev. Mat. Sci. **2**, 1972.
120. *Problemy Vysokotemperaturnoy Sverkhprovodimosti*, Nauka, Moscow, 1977 [*High-Temperature Superconductivity*, ed. by V.L. Ginzburg and D.A. Kirzhnits, Consultants Bureau, New York, 1982].
121. V.L. Ginzburg, Phys. Today **42** (3), 9, 1989.
122. *Proceedings of International Conference on Materials and Mechanisms of Superconductivity. High-Temperature Superconductivity II*, 23–28 July 1989, Stanford, Physica C, **162–164**, pt. 1, 1989.
123. V.L. Ginzburg, Vestn. Akad. Nauk SSSR **11**, 20, 1987 (Compare Vestn. Akad. Nauk SSSR **5**, 7, 1971); Priroda **7**, 16, 1987.
124. V.L. Ginzburg et al., Solid State Comm. **50**, 339, 1984; V.L. Ginzburg, Pis'ma Zh. Eksp. Teor. Fiz. **30**, 345, 1979 [JETP Lett. **30**, 319, 1979]; A.A. Gorbatsevich, Zh. Eksp. Teor. Fiz. **95**, 1467, 1989 [Sov. Phys.–JETP **68**, 847, 1989].
125. V.L. Ginzburg, Dokl. Akad. Nauk SSSR **30**, 397, 1941.
126. V.L. Ginzburg, Dokl. Akad. Nauk SSSR **36**, 9, 1942.
127. V.L. Ginzburg, Akust. Zh. **1**, 31, 1955.
128. V.L. Ginzburg, Zh. Eksp. Teor. Fiz. **14**, 181, 1944.
129. V.L. Ginzburg, Izv. Akad. Nauk SSSR, Ser. Fiz. **11**, 165, 1947.
130. V.L. Ginzburg, Usp. Fiz. Nauk **46**, 348, 1952; Usp. Fiz. Nauk **52**, 494, 1954; Usp. Fiz. Nauk **56**, 146, 1955.
131. V.L. Ginzburg and V.M. Fain, Zh. Eksp. Teor. Fiz. **39**, 1323, 1960 [Sov. Phys.–JETP **12**, 923, 1961].
132. V.L. Ginzburg, D.A. Kirzhnits, and A.A. Lyubushin, Zh. Eksp. Teor. Fiz. **60**, 451, 1971 [Sov. Phys.–JETP **33**, 242, 1971].
133. Yu.S. Barash and V.L. Ginzburg, Pis'ma Zh. Eksp. Teor. Fiz. **15**, 567, 1972 [JETP Lett. **15**, 403, 1972]; Usp. Fiz. Nauk **116**, 5, 1975; Usp. Fiz. Nauk **143**, 345, 1984 [Sov. Phys.–Uspekhi **18**, 305, 1975; Sov. Phys.–Uspekhi **27**, 467, 1984].
134. V.L. Ginzburg, Vestn. Akad. Nauk SSSR **10**, 50, 1990.
135. V.L. Ginzburg, Usp. Fiz. Nauk **163**, 45, 1993 [Sov. Phys.–Uspekhi **36**, 587, 1993].
136. L.V. Mikheev and M.E. Fisher, J. Low Temp. Phys. **90**, 119, 1993.
137. V.L. Ginzburg and G.F. Zharkov, J. Low Temp. Phys. **92**, 25, 1993; Physica C **235–240**, 3129, 1994.
138. V.L. Ginzburg, Contemp. Phys. **33**, 15, 1992; Physica C **1**, 209, 1993.
139. V.L. Ginzburg, Priroda **6**, 6, 1994.
140. V.L. Ginzburg, Usp. Fiz. Nauk **166**, 1033, 1996 [Phys.–Uspekhi **39**, 973, 1996].
141. G.A. Goncharov, Usp. Fiz. Nauk **166**, 1095, 1996 [Phys.–Uspekhi **39**, 1033, 1996].
142. V.L. Ginzburg, Usp. Fiz. Nauk **166**, 169, 1996 [Phys.–Uspekhi **39**, 155, 1996].
143. V.L. Ginzburg, Phys. Rep. **194**, 245, 1990.
144. V.L. Ginzburg, Usp. Fiz. Nauk **167**, 429, 1997 [Phys.–Uspekhi **40**, 407, 1997].

145. V.L. Ginzburg, Usp. Fiz. Nauk **168**, 363, 1998 [Phys.–Uspekhi **41**, 307, 1998].
146. R.M. Arutyunian, V.L. Ginzburg, and G.F. Zharkov, JETP **84**, 1186, 1997; Usp. Fiz. Nauk **167**, 457, 1997 [Phys.–Uspekhi **40**, 435, 1997].
147. V.L. Ginzburg, in *From High-Temperature Superconductivity to Microminiature Refrigeration* (Plenum, New York 1996) p. 277.
148. V.L. Ginzburg and E.G. Maksimov, Supercond.: Phys. Chem. Tech. **5**, 1505, 1992; Physica C **235–240**, 193, 1994.
149. V.L. Ginzburg and Yu.N. Eroshenko, Usp. Fiz. Nauk **165**, 205, 1995; Usp. Fiz. Nauk **166**, 89, 1996 [Phys.–Uspekhi **38**, 195, 1995; Phys.–Uspekhi **39**, 81, 1996].
150. V.L. Ginzburg and E.A. Andryushin, *Superconductivity*, World Scientific, Singapore, 1994.
151. V.L. Ginzburg, *O Fizike i Astrofizike* (*About Physics and Astrophysics*), Kvantum, Moscow, 1995 [V.L. Ginzburg, *The Physics of a Lifetime: Recollections of the Problems and Personalities of 20th Century Physics*, Springer, Berlin 2001)].
152. V.L. Ginzburg, Usp. Fiz. Nauk **172**, 373, 2002 [Phys.-Usp. **45**, 341, 2002].
153. V.L. Ginzburg, *O Nauke, o sebe i o drugikh* (*About Science, Myself and Others*), Fizmatlit, Moscow, 2003 [V.L. Ginzburg, *About Science, Myself and Others*, IOP Publ., Bristol, 2005].
154. V.L. Ginzburg, Voprosy Istorii Estestvoznaniya i Tekhniki (4), 5, 2000.
155. V.L. Ginzburg, Usp. Fiz. Nauk **171**, 1091, 2001 [Phys.-Usp. **44**, 1037, 2001].
156. V.V. Schmidt, *The Physics of Superconductors: Introduction to Fundamentals and Applications*, ed. by P. Muller, A.V. Ustinov, Springer, New York, 1997.
157. E.G. Maksimov, Usp. Fiz. Nauk **174**, 1026, 2004 [Phys.-Usp. **47**, 957, 2004]; Usp. Fiz. Nauk **174**, 1145, 2004 [Phys.-Usp. **47**, 1075, 2004].
158. *Problemy Teoreticheskoi Fiziki i Astrofiziki* [Problems of Theoretical Physics and Astrophysics] (collection of articles, devoted to 70th birthday of V.L. Ginzburg, ed. by L.V. Keldysh and V.Ya. Fainberg, Nauka, Moscow, 1989.
159. *Academician L.I. Mandelshtam. To the Centenary of his Birth*, edited by S.M. Rytov, Nauka, Moscow, 1989.
160. V.L. Ginzburg, *Nedodumannoe, Nedodelannoe...* (*Unconsidered, Unfinished...*) Preprint No. 34, FIAN, Moscow, 2001 (see also [163]).
161. V.L. Ginzburg, Dokl. Akad. Nauk SSSR **156**, 13, 1964 [Doclady Acad. Sci. USSR **9**, 329, 1964].
162. V.L. Ginzburg, L.M. Ozernoi, Zh. Eksp. Teor. Fiz. **47**, 1030, 1964 [Sov. Phys. JETP **20**, 689, 1965].
163. *Seminar: Stat'i i Doklady* (*The Seminar: Papers and Reports*) (compilers: B.M. Bolotovskii and Yu.M. Bruk), Izd. Fiziko-Matematicheskoi Literatury, Moscow, 2006.
164. Usp. Fiz. Nauk. **176**(10), 2006 [Phys.-Usp. **49**(10), 1003, 2006].
165. Usp. Fiz. Nauk **177**(3), 325, 2007 [Phys.-Usp. **50**(3), 293, 2007].
166. Usp. Fiz. Nauk **177**(4), 345, 2007 [Phys.-Usp. **50**(4), 331, 2007].
167. Usp. Fiz. Nauk **177**(5), 553, 2007 [Phys.-Usp. **50**(5), 529, 2007].
168. Journal of Superconductivity and Novel Magnetism **19**(3–5), 2006.
169. Ferroelectrics **354**, 2007.
170. V.L. Ginzburg, *Ob Ateizme, Religii, i Svetskom Gumanizme* (About Atheism, Religion, and Secular Humanism), Moscow, 2008 (in press).

A

Vitaly L. Ginzburg: A Bibliometric Study[1]

by Manuel Cardona[2] and Werner Marx[2]

In this study the citation-based impact of the works of the Russian physicist Vitaly Lazarevich Ginzburg has been analyzed by bibliometric methods. The time-dependent number of mentions of his name (informal citations), the overall reference-based impact (formal citations) and the number of citations of single articles and books have been investigated. The impact time curves (citation history) of his most frequently cited articles and books are presented and discussed. The scientific contributions of the most influential Ginzburg works are analyzed, in particular their impact on recent research.

A.1 Introduction

Vitaly Lazarevich Ginzburg is one of the pioneers of solid state physics and received the physics Nobel Prize in 2003 (together with A.A. Abrikosov and A.J. Leggett) for his groundbreaking work on superconductivity. In addition, he made important contributions to other areas of theoretical physics, particularly to astrophysics and plasma physics. Because of his long scientific life (in contrast to the short scientific life of George Placzek, who was analyzed by the authors in a recent study [1]) Ginzburg is an excellent subject for learning the capabilities and shortcomings of citation analysis. Furthermore, this case study may reveal the coverage of Russian journals in the ISI citation databases and the fraction of citations to the original articles and their translations.

---

[1] Published in *Journal of Superconductivity and Novel Magnetism*, V. 19, No 3–5, July 2006 (@ 2006). For the present edition, the authors have made some amendments.

[2] Max Planck Institute for Solid State Research, D-70569 Stuttgart (Germany).

A.2 The WoS Search Modes

The data presented here is based on the *Science Citation Index* (SCI) accessed through the *Web of Science* (WoS), the search platform provided by Thomson Scientific (the former Institute for Scientific Information; for conciseness we keep referring to the present provider as ISI). The WoS is based on the ISI citation indexes, in particular the SCI, and has probably become the most versatile and user-friendly citation analysis tool. Its competent use, however, requires some experience and awareness of capabilities and pitfalls. The SCI covers a dynamical set of about 6000 source journals, carefully selected by the ISI to be relevant to the progress of science. Beside the classical *Science Citation Index* (SCI) the WoS provides access to the *Social Sciences Citation Index* (SSCI) and the *Arts & Humanities Citation Index* (A & HCI). Recently, the WoS has been extended to cover articles and their citations since 1900 (ISI Century of Science). Hence, the WoS enables the use of bibliometric methods not only in research evaluation, but also in the history of science.

The *Web of Science* provides two search modes: the *General Search* mode and the *Cited Reference Search* mode. The *General Search* mode archives and reveals publications in source journals (the so-called source items) starting in the year 1900. The search results are limited to the articles published in the source journals (which neither include books and popular publications, nor conference proceedings not published in regular journals). Under *General Search* all search fields (topic, title, author, address, source, publication year, etc.) can be searched and combined by logical operators (AND, OR, NOT). This is the standard search mode of the classical literature databases such as the Chemical Abstracts or the Physics Abstracts.

The *Cited Reference Search* mode enables access to all references appeared in source journal articles (whether cited correctly or incorrectly). References to articles published in non-source journals or in books or any other published material (sometimes even unpublished, e.g., theses, internal reports, or even private communications) are thus included. Probably the most useful feature of this search mode is the possibility of finding a measure of the impact of a book or a book article, as reflected in the citations in source journal articles. The references related to source journal articles are linked to the corresponding source items giving access to the full bibliographic data and the abstracts (the latter only since 1991). By this procedure, the source item related references (which appear blue on the monitor) can easily be distinguished from the non-source item or erroneous references (which appear black).

A.3 The Scientific Output

The WoS *General Search* mode reveals 424 articles with V.L. Ginzburg as first author or as a coauthor (among them, three philosophy related articles covered by the SSCI and the A & HCI). These are only source items, i.e.

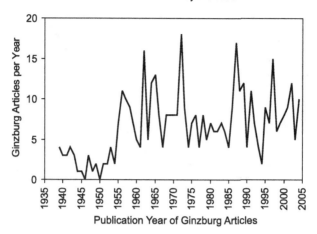

Fig. A.1. Number of Ginzburg articles per year covered by the *Science Citation Index* (SCI) as source items, searched under the *Web of Science* (WoS)

articles published either in ISI source journals or their translations. Books and other non-source article publications are not included. Figure A.1 shows the number of Ginzburg articles appearing in the ISI source journals as function of their publication years. An average number of almost 10 articles per year from 1955 till present is indeed very impressive – the more so as not all publications are included (see below).

Basically, Russian journals are covered by the SCI, provided that they have a sufficiently high impact and/or reputation and a few other editorial requirements established by the ISI. However, the source data is sometimes not consistent, e.g. they differ between the original and the translated items. For example, from 1945 to 1954 and from 1972 to 1996, the articles of the original version of *ZH EKSP TEOR FIZ* appear as source items. From 1955 to 1971, however, the translated version *SOV PHYS JETP* is covered. For reasons unbeknownst to us (see the note added in proof) the complete year 1950 of the translated version is missing. The SCI coverage of another important Russian journal changed in a similar way: the original version *USP FIZ NAUK* ranges from 1955 to 1964 and from 1972 to 2001. The translated version *SOV PHYS USP* fills the space in between, ranging from 1965 to 1971.

For the scientists from the former USSR in the period from 1947 till 1958 there were some difficulties with the publication of articles in English. The very important *Journal of Physics of the USSR* ceased publication around 1947. Probably the already published issues of the English Edition were destroyed. The gap in the flow of Soviet scientific literature to the Western scientific community can be explained, at least in part, as due to the activities of the Soviet government at that time.

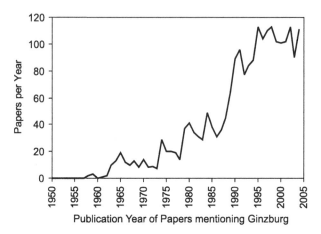

Publication Year of Papers mentioning Ginzburg

Fig. A.2. Time-dependent number of papers mentioning Ginzburg in the titles of articles corresponding to SCI source items (informal citations)

A.4 The Informal Citations

Instead of a full reference to the original work (in terms of a formal citation) sometimes the name of the author or the concept bearing his name is given (e.g., the *Ginzburg–Landau Theory*). In the WoS *General Search* mode, querying for the name of the author as a topic offers the possibility of finding out how many times the author is mentioned as an informal citation. The name 'Ginzburg' has thus been mentioned informally almost 2000 times in article titles since 1939, the year of his first publication. In this count, the abstracts and keywords were not included as search fields. At present, the abstracts are not available prior to 1991 in the SCI and also the inclusion of keywords changes with time. Figure A.2 shows the time dependent number of papers mentioning 'Ginzburg' in the title. Note that the strong increase since 1985 is related to the discovery of high-temperature superconductivity [2]. Notice the small peak in 1965: it will be shown later that it is due to the seminal 1957 paper of A.A. Abrikosov [3] which resulted in the commercial appearance of superconducting magnets in 1962.

Being confined to the titles of the source items, Fig. A.2 gives a somewhat limited picture of Ginzburg's informal citations. It is possible to include not only titles but also abstracts and keywords in the counting of informal citations since 1968 by using the INSPEC database (Physics Abstracts). The result is shown in Fig. A.3, which reveals an increase by a factor of four with respect to Fig. A.2. The curve obtained by counting only mentions in the titles has been included for comparison. Most of the informal citations (ca. 95%) are to the *Ginzburg–Landau Theory* and correspond to Ginzburg's most cited paper (see the next section). Figure A.3 also reveals the strong increase in

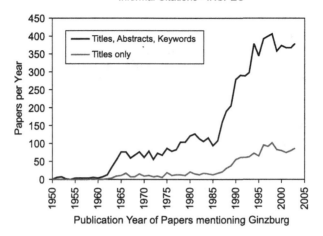

Fig. A.3. Time-dependent number of papers covered by the INSPEC database mentioning 'Ginzburg' in the titles, abstracts and/or keywords. For comparison we have included the curve corresponding to mentions only in the titles

informal citations (most to the *G–L Theory*) related to the discovery of high-temperature superconductivity in 1986.

A.5 The Overall Citation Impact

The influence or significance of scientific articles can be expressed in terms of how often they are noticed and how long they are remembered. Citation numbers are frequently taken as a measure of the resonance or the impact an article, a journal, or a scientist has generated. Figure A.4 shows the time-dependent overall citation impact of Ginzburg's publications. All articles covered by the SCI citing any of the Ginzburg publications (whether published or not in source journals) are included. The citing papers were selected according to their publication years using the WoS *Cited Reference Search* mode, which allows the inclusion of non-source items and incorrect citations. The strong increase between 1960 and 1965 is certainly related to Abrikosov's seminal work [3] and the commercial appearance of superconducting magnets in 1962. Note that the increase of informal citations seen after 1985 in Figs. A.2 and A.3 does not appear in Fig. A.4. This reflects the fact that after a concept, a paper or their authors become household words (e.g., the G–L Theory); the informal citations increase at the expense of the formal ones.

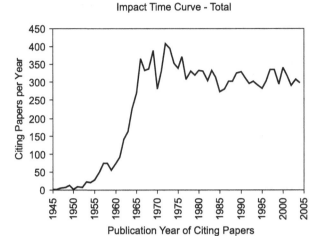

Fig. A.4. Time-dependent number of papers citing Ginzburg publications (books included). Note that the ordinate of this figure corresponds to the number of citing papers which is somewhat lower than the number of citations. One citing paper may include more than one citation (i.e., one article cites two or more Ginzburg articles)

A.6 The Impact of Single Publications

The graph displaying the time-dependent evolution of a single article is sometimes called its citation history. Each article develops its own life span as it is being cited. With time, the citations per year (citation rate) normally evolve following a similar pattern: the citations generally do not increase substantially until one year after publication. They reach a summit after about three years, the peak position depending somewhat on the research discipline. Subsequently, as the articles are displaced by newer ones, their impact decreases, accumulating citations at a lower level. Finally, they are barely cited or forgotten.

This is the pattern found for most articles of ordinary mortals. Some work, however, is well ahead of its time and receives little response when first published. After some time, in some cases after many years or even decades, the field and the community are ready to recognize the work and, correspondingly, acknowledge it by citing. A delayed peak appears in the citation history of such publications, whose growth may even exceed that of the literature in its discipline. Sometimes, such publications are labeled as 'sleeping beauties.'

Several of Ginzburg's articles fall into this category. As an example, we show in Fig. A.5 the citation history of his most highly cited article, entitled 'Theory of Superconductivity', coauthored with L.D. Landau [4]. The onset of the impact of this article appears delayed by about 10 years, compared with the normal citation pattern. Note that the delayed peak, around 1965, must be related to Abrikosov's seminal article which, as already mentioned,

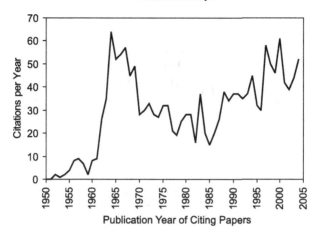

Fig. A.5. Citation history of the most cited Ginzburg article [4] having garnered a total of 1617 citations until now (erroneous citations are not included)

triggered the commercial development of superconducting magnets. The increase that starts around 1985 is related to the discovery of high-temperature superconductivity. At present, this article has a remarkable impact of about 50 citations per year.

Figure A.6 displays the citation history of two other highly cited articles published by Ginzburg. The article on a microscopic theory of ferroelectric materials [5] shows a broad peak around 1975 and has still a remarkable, nearly time-independent impact today. The citations of this paper show no sign of ebbing out, in spite of having been published nearly half a century ago. The citation summit of the article with L.P. Pitaevskii on the theory of superfluidity [6] is delayed by about 40 years!

One should note that the time dependence of citations results from a combination of two different phenomena: the aging of the articles (obsolescence, replacement, oblivion) and the growth of the scientific literature. Since 1900 the number of publications in the natural sciences has increased by a factor between 50 and 100. The proliferation of science implies more potentially citable articles, resulting in an increasing ratio of references per article (reference count). For example, the number of references per article in chemistry and physics increased from less than 10 around 1900 to about 30 at present. As a result, the probability to be cited and therewith the average number of citations per article (average citation rate) almost doubled over the past 50 years [7].

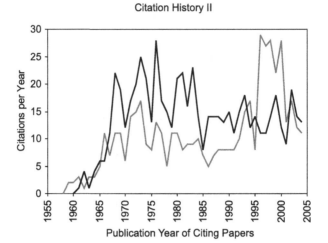

Fig. A.6. Citation history of two highly cited Ginzburg articles. *Light graph*: [5] – 637 citations. *Dark graph*: [6] – 511 citations

A.7 The Citations of Ginzburg's Books

Figure A.7 displays the citation history of the two most highly cited Ginzburg books: *The Origin of Cosmic Rays* (OCR) and *The Propagation of Electromagnetic Waves in Plasmas* (PEW). OCR appeared in Russian in 1963; an English translation followed in 1964. PEW has two Russian editions which appeared in 1960 and 1967. Each of them has an English translation (published in 1964 and 1970, respectively). All editions of OCR have received a total of 1051 citations, whereas those of PEW have been cited a total of 2582 times. It is interesting to note that the yearly citations of the English editions level off at a constant rate (ca. 15 for OCR, ca. 35 for PEW) whereas those of the Russian editions taper off since around 1990, tending to zero in recent years. One may speculate whether this reflects the political and economic upheavals experienced by the USSR/Russia during this period and the decrease in the support of science and, correspondingly, the interest of students; a rather sad conjecture. It may also reflect the loss of weight of Russian as a scientific language.

Besides the two highly cited books whose citation patterns are presented in Fig. A.7, Ginzburg has published a large number of other books, some textbooks and some containing original research. Because of the often large number of versions of a given book and its translations, it is rather difficult to perform a bibliometric analysis in the case of Ginzburg's books. Beside the ones discussed, the most cited of his books is *High-Temperature Superconductivity*, published in Russian by Nauka, Moscow 1977, and coauthored by D.A. Kirzhnits. We have found 15 formal citations to this Russian version in SCI source articles. It was probably a work that was far ahead of its time:

Citation History - Books

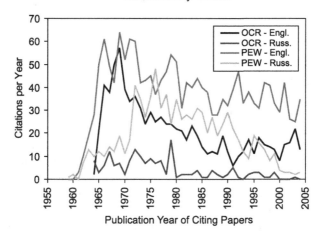

Publication Year of Citing Papers

Fig. A.7. Citation history of the most highly cited Ginzburg books determined by using the *Cited Reference Search* mode of WoS (see rank 1 and 3 of Table A.2). OCR: *The Origin of Cosmic Rays*, PEW: *The Propagation of Electromagnetic Waves in Plasmas*. The Russian and English editions are distinguished

high-temperature superconductivity, as we now know it, was not discovered till 1986.

The first English edition of the book was published by Plenum Press (Consultants Bureau Division) in 1982. It has been cited about 150 times. Its appearance was rather timely: there had been reports in 1978 of superconductivity in CuCl after specific pressure and temperature cycling [8, 9]. Shortly after these reports, similar phenomena were reported for a related semiconductor, CdS [10]. Ginzburg and Kirzhnits discuss this problem in their high-temperature superconductivity book, trying to keep a careful distance to the attribution of superconductivity to the reported phenomena. Ginzburg prefers to call the alleged effect 'superdiamagnetism' [11]. To date this effect has not received sufficient reliable confirmation and is nearly forgotten. The discovery of high-temperature superconductivity in cuprates was made in 1986 [2]. Although no general consensus about its theoretical origin has been reached until present, the book by Ginzburg and Kirzhnits has been rather helpful in sorting out possible mechanisms, as revealed by the considerable number of citations it has garnered since then.

In order to include in this discussion as many of Ginzburg's copious books as possible, we examined the Web for possible availability at used book dealers, a rather effective method which nevertheless may seem unorthodox to other practitioners of bibliometry. We used the link www.abebooks.com, which includes nearly 20,000 such dealers. To our surprise, we found 23 available titles with prices covering the range from $3 (*Physics and Astrophysics: A Selection of Key Problems*) to $303 (*Issues of Intense-Field Quantum Electrodynamics*).

Ginzburg's latest book, of an autobiographical nature (*About Science, Myself and Others*), appeared in 2004. It has not yet had the chance to accumulate many citations.

A.8 The Citation Ranking

The 424 articles by V.L. Ginzburg selected under the WoS *General Search* mode can be ranked by the number of citations using the WoS SORT command. Table A.1 shows the 10 most cited Ginzburg papers that are SCI source articles. This table is flawed concerning two aspects: (1) The citations to books and articles that are not source items are ignored. Hence, the two books and the most cited paper mentioned above are not included. (2) The citations to the original versions and the translations of the Russian journals are not both included. For example, the top paper of Table A.1 corresponds to [6] in Fig. A.6 and garnered 511 instead of 358 citations, as stated in the WoS source item.

We have found, in addition, 23 papers (source items) attributed to V.L. Ginsburg (i.e., misspelled, s instead of z) accounting for a total of 800 citations. Most of the errors have been made in the translation or transliteration (the latter applies largely to French journals).

The WoS *Cited Reference Search* mode allows additionally selecting the citations to books and to article versions (original papers or translations) not being source items. Table A.2 shows a more complete ranking of the Ginzburg publications established by selecting the citations of the corresponding article versions. Note that the overall number of citations of the Russian and the English versions of a given article is less than the sum of both because of double citations. The fraction of double citations is rather different and ranges from less than 1% to up to 30%, depending on the specific journal and the publication date. We also found that V.L. Ginzburg was sometimes cited as V.I. Ginzburg (I instead of L as second initial), which revealed some additional 60 citations.

The total citation-based impact of Ginzburg's publications at the date of search results in the exceptionally large number of 21269 formal citations (book citations are included). Self-citations could be removed but are included here. Excluding self-citations does not affect the significance of such studies considerably.

Another interesting aspect is the time dependence of a researcher's 'creativity', i.e., to time-resolve the number of citations according to when his/her articles were published, rather than when they were cited (i.e., the publication years of the citing papers, as in Figs. A.2–A.7). This kind of graph can be established by what we call the citation vintage diagram. According to the vintage years with their changing quantity and quality (price) a researcher publishes new papers with variable output and impact. With time, the impact not only of the recently published papers but also of the older ones is

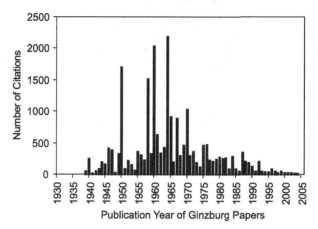

Fig. A.8. Citation vintage diagram (with the number of citations until the present as a function of the publication year of the Ginzburg publications (articles and books). Date of search: 2005-12-01

increasing. However, the increase is uneven – the impact of the newer papers usually grows faster than the impact of the older ones.

In Fig. A.8 we show a bar diagram with the number of citations, collected until the present, to all articles of Ginzburg published in a given year, as a function of the year of publication. In this way, the relative contribution of the author's publication years (and articles) to the overall impact of his works is visualized. E.g., the bar for publication year 1964 represents a total impact of 2177 citations within the time period from 1964 to the present day, to the articles published by Ginzburg in the year 1964 only. This citation pattern is dominated by the peaks resulting from the three most cited papers, published in 1950, 1958, and 1960, as well as the various book editions published in 1960, 1964, and 1970.

The data given in Fig. A.8 is based on the WoS *Cited Reference Search* mode. As already mentioned, this search mode enables counting not only the citations of articles appearing in source journals, but also those of articles published in non-source journals (or any other published material) and in addition the book citations.

A.9 The Hirsch-Index h

A new index (h-index, or h-number) was introduced recently by Jorge E. Hirsch as a measure of the cumulative impact of a person's scientific work within a given discipline [12]. It can be easily obtained under the WoS *General Search* mode, provided that there are either no highly cited namesakes, or

Table A.1. SCI source items of Ginzburg articles searched under WoS and ranked by the numbers of citations (the top 10 out of 424 papers). Date of search: 2005-12-01

1. GINZBURG VL, SYROVATS SI
COSMIC MAGNETOBREMSSTRAHLUNG (SYNCHROTRON RADIATION)
ANNUAL REVIEW OF ASTRONOMY AND ASTROPHYSICS 3: 297& 1965
Times Cited: 358

2. GINZBURG VL, PITAEVSKII LP
ON THE THEORY OF SUPERFLUIDITY
SOVIET PHYSICS JETP-USSR 7 (5): 858–861 1958
Times Cited: 338

3. GINZBURG VL, GUREVICH AV
NELINEINYE YAVLENIYA V PLAZME, NAKHODYASHCHEISYA
V PEREMENNOM ELEKTROMAGNITNOM POLE
USPEKHI FIZICHESKIKH NAUK 70 (2): 201–246 1960
Times Cited: 260

4. GINZBURG VL
ON SURFACE SUPERCONDUCTIVITY
PHYSICS LETTERS 13 (2): 101–102 1964
Times Cited: 124

5. GINZBURG VL
ON THE DESTRUCTION AND THE ONSET OF SUPERCONDUCTIVITY
IN A MAGNETIC FIELD
SOVIET PHYSICS JETP-USSR 7 (1): 78–87 1958
Times Cited: 123

6. GINZBURG VL, PTUSKIN VS
ORIGIN OF COSMIC-RAYS:
SOME PROBLEMS IN HIGH-ENERGY ASTROPHYSICS
REVIEWS OF MODERN PHYSICS 48 (2): 161–189 1976
Times Cited: 120

7. GINZBURG VL
FERROMAGNETIC SUPERCONDUCTORS
SOVIET PHYSICS JETP-USSR 4 (2): 153–160 1957
Times Cited: 118

8. GINZBURG VL, SYROVATS SI
DEVELOPMENTS IN THEORY OF SYNCHROTRON RADIATION
AND ITS REABSORPTION
ANNUAL REVIEW OF ASTRONOMY AND ASTROPHYSICS 7: 375& 1969
Times Cited: 112

9. GINZBURG VL
PROBLEM OF HIGH-TEMPERATURE SUPERCONDUCTIVITY 2.
SOVIET PHYSICS USPEKHI-USSR 13 (3): 335& 1970
Times Cited: 111

10. GINZBURG VL
SCATTERING OF LIGHT NEAR PHASE TRANSITION POINTS IN SOLIDS
USPEKHI FIZICHESKIKH NAUK 77 (4): 621–638 1962
Times Cited: 92

Table A.2. Citation ranking of the Ginzburg papers with more than 100 citations based on the WoS *Cited Reference Search* mode. The overall number of citations and the citations of the corresponding original papers and their translations are given. The total number includes additional citations due to reference variants: the original source title is combined with the volume and page number of the translation or vice versa; sometimes the volume and page number of the original paper and the translation are combined cross-over. When establishing the total numbers, simultaneous citations of original papers and translations (double citations) are considered. The erroneous citations (incorrect numerical reference data) are not included. Light font: SCI source items. Date of search: 2005-12-01. *Column 1*: Consecutive numbering; *Column 2*: Total number of citations at the date of search; *Column 3*: Separate number of citations of original papers and translations from; *Column 4*: Short form of the cited publications

1	2584	848	GINZBURG VL	1960, 1967			RASPROSTRANENIE ELEKTROMAG
		1749	GINZBURG VL	1964, 1970			PROPAGATION ELECTROMAG WAVES
2	1617		GINZBURG VL	1950	V20	P1064	ZH EKSP TEOR FIZ
3	1047	169	GINZBURG VL	1963			PROISKHOZHDENIE KOSMIC
		884	GINZBURG VL	1964			ORIGIN COSMIC RAYS
4	637	301	GINZBURG VL	1960	V2	P2031	FIZ TVERD TELA
		414	GINZBURG VL	1960	V2	P1824	SOV PHYS-SOLID STA
5	511	187	GINZBURG VL	1958	V34	P1240	ZH EKSP TEOR FIZ
		338	GINZBURG VL	1958	V7	P858	SOV PHYS JETP
6	418	260	GINZBURG VL	1960	V70	P201	USP FIZ NAUK (I)
		190	GINZBURG VL	1960	V3	P115	SOV PHYS USP (I)
		73	GINZBURG VL	1960	V70	P393	USP FIZ NAUK (II)
		78	GINZBURG VL	1960	V3	P175	SOV PHYS USP (II)
7	358		GINZBURG VL	1965	V3	P297	ANNU REV ASTRON ASTR
8	305		GINZBURG VL	1946	V16	P15	ZH EKSP TEOR FIZ
9	169		GINZBURG VL	1958	V2	P653	SOV ASTRON
10	164	79	GINZBURG VL	1970	V101	P185	USP FIZ NAUK
		111	GINZBURG VL	1970	V13	P335	SOV PHYS USP
11	162	66	GINZBURG VL	1976	V120	P153	USP FIZ NAUK
		104	GINZBURG VL	1976	V19	P773	SOV PHYS USP
12	152	72	GINZBURG VL	1958	V34	P113	ZH EKSP TEOR FIZ
		123	GINZBURG VL	1958	V7	P78	SOV PHYS JETP
13	152		GINZBURG VL	1979			THEORETICAL PHYSICS
14	144		GINZBURG VL	1982	V2		HIGH-TEMPERATURE SUPERCOND
15	141		GINZBURG VL	1949	V38	P490	USP FIZ NAUK
16	132		GINZBURG VL	1977			PROBLEMA VYSOKOTEMPE
17	124		GINZBURG VL	1964	V13	P101	PHYS LETT
18	121		GINZBURG VL	1970	P255		VOLNY MAGNITOAKTIVNO
19	120		GINZBURG VL	1976	V48	P161	REV MOD PHYS
20	118		GINZBURG VL	1957	V4	P153	SOV PHYS JETP
21	112		GINZBURG VL	1969	V7	P375	ANNU REV ASTRON ASTR

Table A.3. Russian physics Nobel Laureates with the number of citations of papers published in SCI source journals and the corresponding h-numbers. Please note that the data does not reveal the complete citation impact. Date of search: 2005-11-25

Author	# citations	h-index
A.A. Abrikosov (1928–)	5573	30
Z.I. Alferov (1930–)	7416	40
N.G. Basov (1922–2001)	4607	33
P.A. Cherenkov (1904–1990)	59	4
I.M. Frank (1908–1990)	469	10
V.L. Ginzburg (1916–)	5732	40
P.L. Kapits(z)a (1894–1984)	1011	13
L.(D.) Landau (1908–1968)	5766	31
A.M. Prokhorov (1916–2002)	8866	37
I.(Y.) Tamm (1895–1971)	6493	41

they can be easily removed. The h-index is simply defined as the number of articles in source journals that have had h citations or more. The index increases with the age of the scientist and depends on his specific research field. The h-index reflects a researcher's contribution based on a broad body of publications rather than based on a few high-impact papers. This avoids an overestimation of a single or a few highly-cited papers, sometimes being methodological contributions or reviews. The h-index favors researchers who consistently produce influential papers.

The reference citations and the h-index numbers of Table A.3 were determined on the basis of the WoS *General Search* mode. In this mode only the citations of the papers covered by the SCI source journals are included. As a result of this procedure, a considerable loss of citations may occur, especially in the case of Russian authors. For example, the h-index of Ginzburg raises from 40 to above 60 when the citations to his books and the many articles which are not SCI source items are taken into account. Furthermore, the impact of early papers, and thereby the overall impact of pioneers like Kapitza or Cherenkov, are highly underestimated. The increasing number of citable papers within the last century results in increasing average citation rates. Hence, the citation numbers (and the h-numbers) of scientists who lived at different times are hardly comparable.

A.10 Concluding Remarks

We have carried out a detailed analysis of the citations of Ginzburg's articles covering his whole scientific life (ca. 65 years). In this analysis we have considered the citations of SCI source articles as well as those of any other publications (e.g., books), provided they were cited in source articles. So-called informal citations, involving mentions of Ginzburg's name without a specific formal citation, have also been investigated. Our studies place Ginzburg among the most prominent and influential Russian physicists. According to the Hirsch number (h-index = 40) Ginzburg is one of the most influential Russian Nobel Laureates. According to the total number of citations, he occupies the 5th place.

The Thomson/ISI citation indexes are not at all complete, but are still the main source of impact data for scientists all over the world. At present, the WoS is the only citation database reaching sufficiently far back to enable studies of the history of science. In the present case study we have attempted to demonstrate not only the citation-based impact of Ginzburg's outstanding work, but also the shortcomings of the WoS, in particular concerning (early) Russian literature in a case study. This analysis is of particular interest in revealing the effects of the language (Russian vs. English) on the impact of a publication as measured by the citations received. Publications in English are cited considerably more often than their Russian counterparts even though the latter correspond to the original work, whereas the former are merely translations.

Note Added in Proof

After the manuscript was completed, A. Wittlin pointed out to us the possible reason why the complete 1950 issues of the ZhETF are missing in the Western literature databases. This reason can be found in the closing chapter of R.D. Parks's book on superconductivity (Marcel Dekker, New York, Vol. 2, p. 1347, 1969), written by P.W. Anderson. The year 1950 marks the early days of the Cold War as well as the concomitant beginning of McCarthyism in the USA. In those witch-hunting days, Soviet publications, including scientific journals, were either formally or informally banned. Anderson suggests that issues arriving to the USA may have been dumped straight into the harbors. Many physicists working on superconductivity in the West did not become aware of [4] till the publication of Abrikosov's seminal paper in 1957 [3].

V.M. Agranovich mentioned to us that the two books he coauthored with Ginzburg are not included in this article. The reason is that the WoS searching routine for books (*Cited Reference Search* mode) only reacts to the first author of multiple author books. The Agranovich-Ginzburg books are:

V.M. Agranovich, V.L. Ginzburg, *Spatial Dispersion in Crystal Optics and the Theory of Excitons*, Wiley (1966) Times Cited: 534 (Russ. 1965 Edition incl.)

V.M. Agranovich, V.L. Ginzburg, *Crystal Optics with Spatial Dispersion and Exitons*, Springer (1984) Times Cited: 795 (Russ. 1979 Edition incl.)

References

1. M. Cardona, W. Marx, Georg(e) Placzek: a bibliometric study of his scientific production and its impact. URL: http://arxiv.org/abs/physics/0601113
2. J.G. Bednorz, K. A. Müler, *Possible high-TC superconductivity in the BA-LA-CU-O system.* Z. Phys. B Con. Mat. **64**, 2, 189–193, 1986 Times Cited: 7500.
3. A.A. Abrikosov, *On the magnetic properties of superconductors of the second group.* Sov. Phys. JETP-USSR **5**, 6, 1174–1183, 1957 Times Cited: 2000.
4. V.L. Ginzburg, L.D. Landau, *Theory of superconductivity.* Zh. Eksp. Teor. Fiz. **20**, 1064–1082, 1950 Times Cited: 1617.
5. V.L. Ginzburg, *Some remarks on phase transitions of the second kind and the microscopic theory of ferroelectric materials.* Sov. Phys. Solid State **2**, 1824, 1960 Times Cited: 637.
6. V.L. Ginzburg, L.P. Pitaevskii, *On the theory of superfluidity.* Sov. Phus. JETP-USSR **7**, 5, 858–861, 1958 Times Cited: 511.
7. E. Garfield, *Random thoughts on citationology – its theory and practice.* Scientometrics **43**, 1, 69–76 Sep 1998 Times Cited: 14.
8. N.B. Brandt, S.V. Kuvshinnikov, A.P. Rusakov, et al., *Anomalous diamagnetism (high-temperature meissner effect) in compound cucl.* JETP Lett. **27**, 1, 33–38, 1978 Times Cited: 100.
9. C.W. Chu, A.P. Rusakov, S. Huang, et al., *Anomalies in cuprous chloride* Phys. Rev. B **18**, 5, 2116–2123, 1978 Times Cited: 110.
10. E. Brown, C.G. Homan, R.K. Maccrone, *Flux exclusion in CDS at 77-K – superconductivity at high-temperatures.* Phys. Rev. Lett. **45**, 6, 478–481, 1980 Times Cited: 50.
11. V.L. Ginzburg, A.A. Gorbatsevich, Y.V. Kopayev, et al., *On the problem of superdiamagnetism.* Solid State Commun. **50**, 4, 339–343, 1984 Times Cited: 38.
12. J.E. Hirsch, Proc. Nat. Acad. Sciences **102**, 16569–16572, 2005. *In a recent web page PNAS has announced that the article by Hirsch has been the most often downloaded article from the PNAS server during November of 2005.*